Advanced Drilling Engineering: Principles and Designs

Advanced Drilling Engineering: Principles and Designs

Editor

Manoj Karkare

scitus
academics

Advanced Drilling Engineering: Principles and Designs

Edited by **Manoj Karkare**

Printed in 2017

ISBN: 978-1-68117-324-5

Library of Congress Control Number: 2015939237

© 2016 by
SCITUS Academics LLC,
616, Corporate Way, Suite 2, 4766,
Valley Cottage, NY 10989

www.scitusacademics.com

Notice

Contents

vi

Preface

This book presents the fundamental principles of drilling engineering, with the primary objective of making a good well using data that can be properly evaluated through geology, reservoir engineering, and management. It is written to assist the geologist, drilling engineer, reservoir engineer, and manager in performing their assignments. The topics are introduced at a level that should give a good basic understanding of the subject and encourage further investigation of specialized interests. Many organizations have separate departments, each per forming certain functions that can be done by several methods. The reentering of old areas, as the industry is doing today, particularly emphasizes the necessity of good holes, logs, casing design, and cement job. Proper planning and coordination can eliminate many mistakes, and I hope the topics discussed in this book will play a small part in the drilling of better wells. This book was developed using notes, comments, and ideas from a course I teach called "Drilling Engineering with Offshore Considerations."

Editor

Chapter 1

Design of Telerobotic Drilling Control System with Haptic Feedback

Faraz Shah and Ilia G. Polushin

Department of Electrical and Computer Engineering, Western University, London, ON, Canada N6A 5B9

ABSTRACT

The paper deals with the design of control algorithms for virtual reality based telerobotic system with haptic feedback that allows for the remote control of the vertical drilling operation. The human operator controls the vertical penetration velocity using a haptic device while simultaneously receiving the haptic feedback from the locally implemented virtual environment. The virtual environment is rendered as a virtual spring with stiffness updated based on the estimate of the stiffness of the rock currently being cut. Based on the existing mathematical models of drill string/drive systems and rock cutting/penetration process, a robust servo controller is designed which guarantees the tracking of the reference vertical penetration velocity of the drill bit. A scheme for on-line estimation of the rock intrinsic specific energy is implemented. Simulations of the proposed control and parameter estimation algorithms have been conducted; consequently, the overall telerobotic drilling system with a human operator controlling the process using PHANTOM Omni haptic device is tested experimentally, where the drilling process is simulated in real time in virtual environment.

INTRODUCTION

Drilling a borehole is a common method for extracting oil, gas, and natural resources from beneath the surface of the earth. Conventional oil well drilling has made significant progress over recent years, and currently is one of the most automated processes in the oil and gas industry. However, there are still some fundamental challenges associated with the drilling. One of the challenges is the choice of vertical penetration velocity of the drill bit. For efficient drilling operation, this velocity must depend upon the type of rock beds drilled. In particular, the velocity must be adjusted when mechanical characteristics of rock strata change. Often, it is difficult to estimate in real time the relative position of the drill bit with respect to different rock layers and, therefore, hard to predict the mechanical characteristics of the rock formations.

The goal of this research is to design a telerobotic system with haptic feedback for control of the drilling process. Telerobotics for drilling well is a relatively novel idea, and it is substantial endeavor to automate one of the fundamental processes in the extraction of energy and resources. As telerobotics is integrated with drilling, it can greatly decrease the number of people working and monitoring operation on the site. This, in particular, can reduce the work site hazards. Also, telerobotics can bring actual analysis of in situ conditions (underground drilling environment) in real time to the human operator that works remotely, where (s) he will be able to monitor the current drilling conditions and, in particular, promptly enforce changes in the vertical speed of penetration of the drill bit in the oil well. Real-time control and optimization of the drilling speed are crucial for today's drilling industry, as it can reduce time and immense cost associated with the drilling an oil well. Introduction of haptic feedback would allow the human operator to feel the changes in mechanical characteristics of the rock and adjust the vertical velocity of penetration accordingly.

In this paper, we address the problem of design of control algorithms for virtual reality based telerobotic system with haptic feedback that allows for the remote control of the vertical drilling operation. Based on a simplified mathematical model of the drilling process, control algorithms are designed which allow to achieve a desired rate of the vertical penetration, regardless of the mechanical properties of the rock. The control design includes an online parameter estimator of the intrinsic specific energy which is a parameter that describes the hardness of the rock. All these algorithms are consequently used in the design of a telerobotic drilling system with virtual environment-based haptic feedback that allows the human operator to feel the stiffness of the rock in contact with the drill bit. Simulations and semiexperimental results are performed which confirm the validity of the theoretical developments.

The potential application domain of this research is not limited to onshore/offshore oil well drilling, but the same principles can be applied, in particular, to different types of mining robots [1], telerobotic systems for dredging and mining ocean [2–5], surgical drilling [6], and telerobotic systems for drilling the extraterrestrial terrain to discover and research the minerals and composition beneath [5, 7].

The structure of the paper is as follows. In Section 2, a mathematical model of the drilling process is derived which is subsequently used for the control design. Section 3 deals with the design of control algorithms for rotational and translational motion of the drilling systems, as well as the design of an online parameter estimator of the intrinsic specific energy of the rock. In Section 4, the structure of a telerobotic drilling system is described and the corresponding experimental results are presented. Finally, in Section 5, some conclusions are given and possible future directions are formulated.

MATHEMATICAL MODEL OF DRILLING SYSTEM

In this section, mathematical models that describe the drilling system are presented. Specifically, the mathematical model of drill string and drive system is described in Section 2.1, while the model of rock cutting and penetration is the subject of Section 2.2.

Mathematical Model of the Drill String and Drive System

The drill string is the assembly of rotating pipes which are responsible for transmitting rotation and weight to the bit and bridge up a connection between the bottom hole tools [8]. The components of a drill string along with drill pipes and the bottom hole assembly (BHA) are shown in Figure 1. A number of simplified mathematical models for drill string and drive systems were proposed in the literature, such as [9–12]. The model used in our work was developed in [9]. This model describes the drill string as a simple torsional pendulum, where the drill pipes are represented as torsional springs and the bottom hole assembly is described as a rigid body with inertia. The model is based on the following simplifying assumptions.

- The bottom hole assembly and the drill bits behave like rigid bodies.
- The moment of inertia of the drill pipe is considered to be small in comparison with the moments of inertia of the bottom hole assembly and the rotary table and, therefore, neglected.
- The nonzero time propagation of the torsional force disturbances along the drill string is neglected. The forces assume to propagate instantaneously along the drill string.

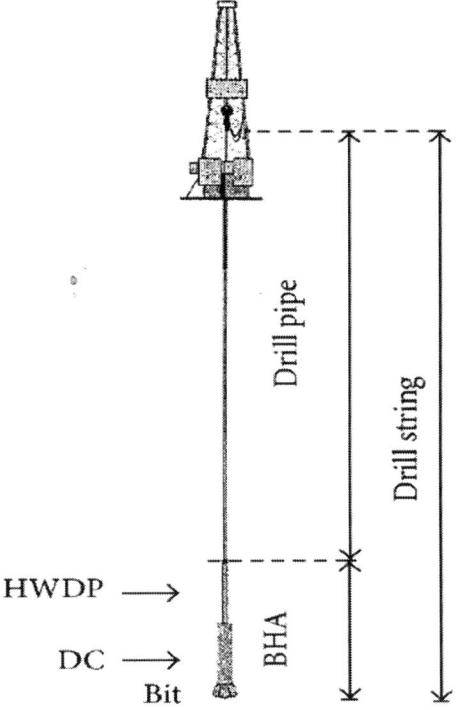

Figure 1: Drill string components [8].

Under the above described assumptions 1–3, the whole drill string and drive system with equivalent electro-mechanical components can be represented by its structural diagram shown in Figure 2. This system is described by the following mathematical model [9]. First, the motion of the drill string is described by the following equation:

$$J_1\ddot{\phi}_1 + c_1\dot{\phi}_1 + k\left(\phi_1 - \phi_2\right) - T = 0.$$

(1)

Here, ϕ_1 is the angular displacement of bit and drill collars (BHA), ϕ_2 is the angular displacement of the rotary table, J_1 is the equivalent moment of inertia of the collars (BHA) and the drill pipes, coefficient c_1 represents equivalent viscous damping, k is the equivalent torsional stiffness of the drill pipes, and T is the torque-on-bit (TOB) generated during the rock cutting process (see Section 2.2 below). The dynamics of the rotary table anddrive systemisdescribedby the following equation:

$$J_2\ddot{\phi}_2 + c_2\dot{\phi}_2 - k\left(\phi_1 - \phi_2\right) - nT_m = 0,$$

(2)

where J_2 is combined moment of inertia of the rotary table and of the rotor of the electric motor coupled together with a gearbox that has 1: n gear ratio, c_2 is aggregated damping of all the components of the drive system, and T_m is the motor torque. Finally, the electric motor is described by the following equations:

$$L\dot{I} + RI + V_b - V = 0, \qquad V_b = K\dot{\phi}_3 = Kn\dot{\phi}_2,$$

$$T_m = KI,$$

(3)

where I is the armature current, L is an equivalent armature inductance, R is an equivalent armature resistance, V_b is the back emf, V is the armature voltage, $\dot{\phi}_3$ is the rotor angular velocity, and K is a constant that depends upon the motor characteristics..

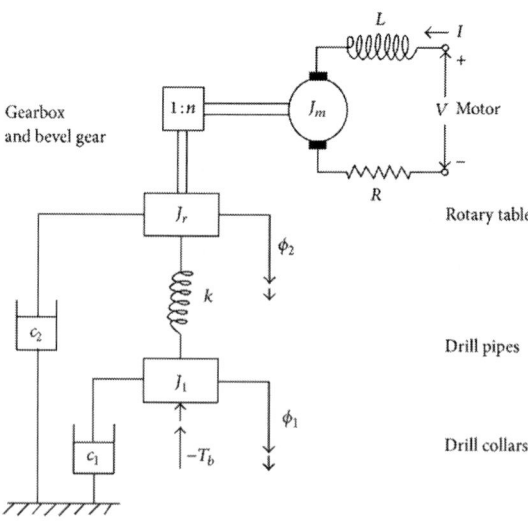

Figure 2: Representation of drill string/drive system with mechanical and electrical components [9].

By combining all the above equations, the complete drill string/drive system can be written in the following state space form:

$$
\begin{bmatrix} \dot{\phi}_1 \\ \dot{\omega}_1 \\ \dot{\phi}_2 \\ \dot{\omega}_2 \\ \dot{I} \end{bmatrix} = \begin{bmatrix} 0 & 1 & 0 & 0 & 0 \\ \dfrac{-k}{J_1} & \dfrac{-c_1}{J_1} & \dfrac{k}{J_1} & 0 & 0 \\ 0 & 0 & 0 & 1 & 0 \\ \dfrac{k}{J_2} & 0 & \dfrac{-k}{J_2} & \dfrac{-c_2}{J_2} & \dfrac{Kn}{J_2} \\ 0 & 0 & 0 & \dfrac{-Kn}{L} & \dfrac{-R}{L} \end{bmatrix} \begin{bmatrix} \phi_1 \\ \omega_1 \\ \phi_2 \\ \omega_2 \\ I \end{bmatrix} + \begin{bmatrix} 0 \\ \dfrac{-T}{J_1} \\ 0 \\ 0 \\ \dfrac{V}{L} \end{bmatrix}
$$

$$(4)$$

Here, ω_1 and ω_2 are the angular velocities of the drill bit and the rotary table, respectively. Equation (4) is valid when the drill bit rotational velocity is greater than zero, that is, $\omega_1 > 0$. In order to reduce the number of equations, a variable ϕ is introduced as the difference of ϕ_2 and ϕ_1. In this case, the original system can be rewritten in the following reduced state space form:

$$
\begin{bmatrix} \dot{\omega}_1 \\ \dot{\phi} \\ \dot{\omega}_2 \\ \dot{I} \end{bmatrix} = \begin{bmatrix} \dfrac{-c_1}{J_1} & \dfrac{k}{J_1} & 0 & 0 \\ -1 & 0 & 1 & 0 \\ 0 & \dfrac{-k}{J_2} & \dfrac{-c_2}{J_2} & \dfrac{Kn}{J_2} \\ 0 & 0 & \dfrac{-Kn}{L} & \dfrac{-R}{L} \end{bmatrix} \begin{bmatrix} \omega_1 \\ \phi \\ \omega_2 \\ I \end{bmatrix} + \begin{bmatrix} \dfrac{-T}{J_1} \\ 0 \\ 0 \\ \dfrac{V}{L} \end{bmatrix}.
$$

$$(5)$$

Equation (5) defines the reduced order model of the drill string and drive system. The model (5) is used for the control design below.

Rock Cutting and Vertical Penetration Models

Astandard drill bit usually exhibits two kinds of motions: rotational along its axis of rotation and vertical motion while penetrating through the rocks. As described in [13], in the normal mode of operation of the drill bit, the bit rotational velocity ω is parallel to its axis of rotation, and the penetration velocity V is directed vertically straight through the rocks. Similarly, the weight-on-bitWacts in the vertical direction and the torqueon- bit T is applied in parallel to the direction of rotation of drill bit. The cutting components of the weight-on-bit and torque-on-bit depend on the radius of PDC drill bit a, intrinsic specific energy ϵ, a parameter $\zeta > 0$ which represents the ratio of the vertical force to the horizontal force between rock and cutter contact surfaces, and the depth of cut d. The depth of cut d plays significant role in the equations to follow that describe the cutting components of the torque-on-bit T and the weight-on-bit W. The equations for these two cutting components are as follows [13]:

$$T^c = \frac{1}{2}a^2\epsilon d,$$

(6)

$$W^c = a\zeta\epsilon d.$$

(7)

In this work, the system is developed under simplifying assumption that the friction effects are negligible. In this case, both variables $T \approx T^c$ and $W \approx W^c$ are proportional to the depth of cut d, according to (6) and (7). As illustrated in Figure 3, the depth of cut d is the thickness of rock ridge in front of the blade. It is assumed that the drill bit has n number of identical blades, and the difference of angular positions of these two successive blades is $(2\pi/n)$. In this case, d is the combined depth of cut of all n blades in each revolution of drill bit, according to the formula

$$d(t) := nd_n(t),$$

(8)

where d_n is the depth of cut of each blade. The depth of cut for each blade is in turn defined according to the formula

$$d_n(t) := U(t) - U(t - t_n),$$

(9)

where $U(t)$ and $U(t - t_n)$ are the vertical positions of the drill bit at current time instant t and a certain previous instant $t - t_n$, respectively [10, 11]. The delay t_n in the above formula is exactly the time that is required for the drill bit to rotate by an angle $2\pi/n$ to achieve its current angular position $\phi_1(t)$; in other words, it also satisfies the following equation:

$$\phi(t) - \phi(t - t_n) = \frac{2\pi}{n}.$$

(10)

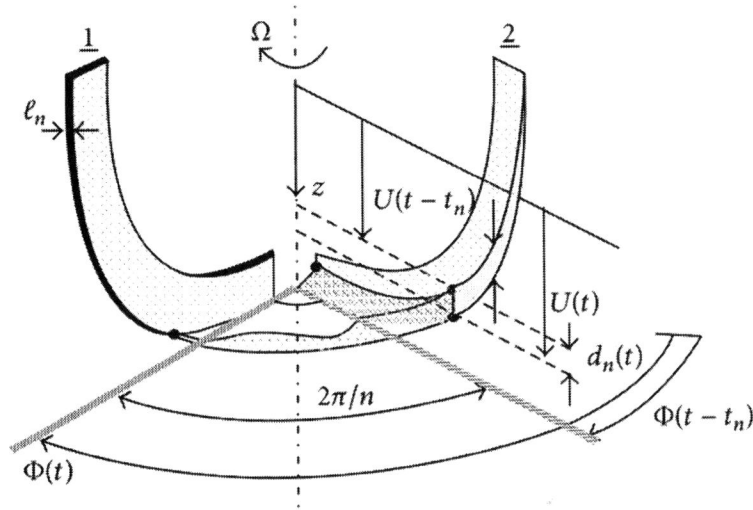

Figure 3: Section of the bottom hole profile located between two successive blades [10].

Using (9) and (10) for calculating (t) would significantly complicate the control design. In this work, we simplify this problem by assuming that both the vertical and angular velocities change slowly; specifically, it is assumed that both $V(\tau) \equiv U(\tau)$ and $\omega_1(\tau) \equiv \phi_1(\tau)$ are approximately constant during each period $\tau \in [t-t_n, t]$. Using this assumptions, (9) and (10) can be rewritten as follows:

$$d(t) \approx n \cdot v(t) \cdot t_n,$$

(11)

$$\omega_1\left(t\right)\cdot t_n \approx \frac{2\pi}{n}.$$

(12)

Combining (11), (12), and assuming $\omega_1(t) \neq 0$,, one gets the following approximate expression for d(t):

$$d\left(t\right) \approx \frac{2\pi \cdot t\left(t\right)}{\omega_1\left(t\right)}.$$

(13)

The above formula has a singularity at $\omega_1(t) = 0$. To remove this singularity, note that the drilling occurs when both $\omega_1(t) > 0$ and V(t) > 0. On the contrary, $\omega_1(t) \leq 0$, the drill bits do not cut the rock and therefore $d(t) \equiv 0$ in this case. Based on the above considerations, one can approximately define the depth of cut according to the formula

$$d\left(t\right) \approx \frac{2\pi \cdot v\left(t\right)}{\max\left\{\omega_1\left(t\right),\epsilon_0\right\}},$$

(14)

where $\epsilon_0 > 0$ is sufficiently small positive constant. The formula (14) does not have singularity at $\omega_1(t) = 0$; it will be occasionally used for calculations of $d(t)$ instead of (13) in the cases where avoiding singularity is important (in simulations, etc.)..

Finally, the vertical motion of the drill bit is described by the following equation [12]:

$$M\frac{dv}{dt} = W_s - W - H_0 - K_f v.$$

(15)

Here, V is the vertical velocity of the drill bit, M is the combined mass of the drill string and BHA,$_0$ is the constant upward force applied from the top of drilling rig, and W_s is the submerged weight of the drill string and Bottom Hole Assembly (BHA). In this model, it is assumed that W_s and H_0 to be constants and defined their difference with another constantW_0 such that$W_0 = W_s - H_0$. Also, W is the applied weight on bit from the interaction of rock defined by (7), and $K_f > 0$ is the coefficient of viscous friction.

CONTROLLER DESIGN

The block diagram of the overall drilling system is shown in Figure 4. As it can be seen from this figure, the block diagram has a complex structure and consists of several interconnected subsystems. Specifically, the vertical motion subsystem is described by (15); the output of this subsystem is the vertical velocity of penetration V(t). The subsystem that represents the rotational motion is described by (5); this subsystem has one control input which is the armature voltage $V(t)$ and one output which is the angular velocity of the drill bits $\omega_1(t)$. Both V(t) and $\omega_1(t)$ are the inputs of the nonlinear static block that represents the cutting process; this subsystem generates the depth of cut $d(t)$ according to (13). Both the torque-on-bit T and weight-on-bit W are proportional to d; they are fed back to rotational motion and vertical motion subsystems, respectively.

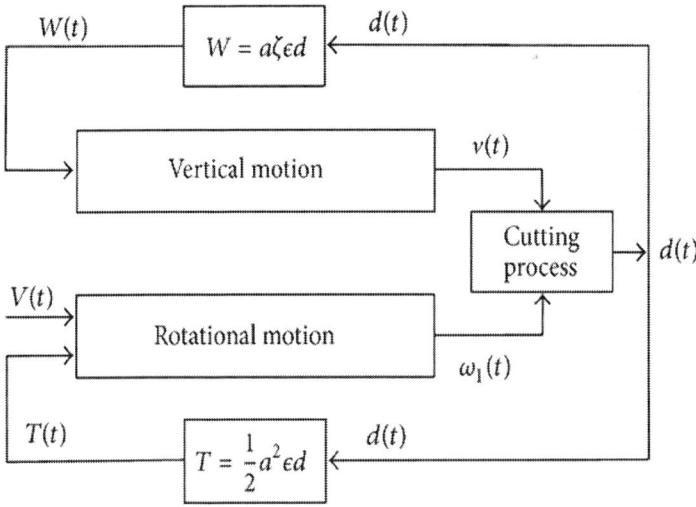

Figure 4: The block diagram of the drilling system.

Our goal is to design a control system that maintains a desired rate of drilling. Specifically, we are looking for the control algorithmfor the armature voltage*V*thatwould guarantee that the velocity of the vertical penetration V(*t*) tends asymptotically to an arbitrary positive desired value $V_{ref} > 0$. We start designing a control algorithm by considering the equation of vertical motion (15) in some detail..

Control of the Vertical Motion of a Drill Bit

The vertical motion of the drilling system is described by (15). For convenience, this equation is rewritten below in a slightly modified form, as follows:

$$\dot{v} = -\frac{K_f}{M} v - \frac{(W_s - H_0)}{M} - \frac{W}{M}.$$

(16)

The idea of the controller developed in this work is to use the weight-on-bit W as the control input to the vertical motion subsystem (16). More specifically, combining formulas (7) and (13), one get the following expression for W:

$$W = a\check{\zeta}\epsilon\frac{2\pi}{\omega_1}v,$$

(17)

which essentially indicates that W is proportional to the vertical velocity v(t) and inversely proportional to the angular velocity of the rotational motion $\omega_1(t)$. Substituting the last formula into (16), one gets

$$= \frac{W_s - H_0}{M} - \frac{1}{M}\left(a\check{\zeta}\epsilon\frac{2\pi}{\omega_1} + K_f\right),$$

(18)

Equation (18) is a linear differential equation with respect to v which, assuming $\omega_1 > 0$, has one stable equilibrium $v = v_0$ defined by the formula

$$\frac{W_s - H_0}{M} - \frac{1}{M}\left(a\zeta\epsilon\frac{2\pi}{\omega_1} + K_f\right)v_0 = 0.$$

(19)

Solving the above equation with respect to v_0, one gets

$$v_0 = \frac{W_s - H_0}{\left(a\zeta\epsilon\left(2\pi/\omega_1\right) + K_f\right)}.$$

(20)

The above equation (20) indicates that the location of the stable equilibrium $V = V_0$ of the vertical motion subsystem (16) can be controlled if one can control the rotational velocity ω_1. Specifically, (20) defines one-to-one correspondence between $\omega 1$ from the range $(0, +\infty)$ and V_0 from the range $(0, (W_s-H_0)/K_f)$. In particular, for any given $V_{ref} \in (0, (W_s - H_0)/K_f)$, there exists an unique $\omega ref \in (0,+\infty)$ such that if the angular velocity satisfies $\omega_1(t) \equiv \omega_{ref}$, then V_{ref} is a globally exponentially stable equilibrium of the translational dynamics (16). For a given $V_{ref} \in (0,(W_s - H_0)/K_f)$, the corresponding ωref can be found using formula (20), as follows:

$$\omega_{ref} = \frac{2\pi a\zeta\epsilon}{\left((W_s - H_0)/v_{ref}\right) - K_f}.$$

(21)

Therefore, the control goal of stabilization of the vertical penetration velocity $V(t) \to V_{ref}$ can be achieved by designing a controller for rotational motion that guarantees a sufficiently fast convergence of $\omega_1(t) \to \omega_{ref}$. The design of such a controlled is presented in the next section..

Stabilization of the Angular Velocity of the Drilling System

The rotational dynamics of the drilling system together with the electric drive are described by (5), which is repeated below for convenience,

$$
\begin{bmatrix} \dot{\omega}_1 \\ \dot{\phi} \\ \dot{\omega}_2 \\ \dot{I} \end{bmatrix} = \begin{bmatrix} \dfrac{-c_1}{J_1} & \dfrac{k}{J_1} & 0 & 0 \\ -1 & 0 & 1 & 0 \\ 0 & \dfrac{-k}{J_2} & \dfrac{-c_2}{J_2} & \dfrac{Kn}{J_2} \\ 0 & 0 & \dfrac{-Kn}{L} & \dfrac{-R}{L} \end{bmatrix} \begin{bmatrix} \omega_1 \\ \phi \\ \omega_2 \\ I \end{bmatrix}
$$

$$
+ \begin{bmatrix} 0 \\ 0 \\ 0 \\ \dfrac{1}{L} \end{bmatrix} V + \begin{bmatrix} \dfrac{-1}{J_1} \\ 0 \\ 0 \\ 0 \end{bmatrix} T.
$$

(22)

The above system has one control input which is the armature voltage of the electric drive V and one disturbance input which is the torque-on-bit T. Our objective in this section is to design a control law for T which would track the reference angular velocity of the drill ω_1 ω_{ref} while rejecting the disturbance T.

To solve the control problem formulated above, one can use the approach to feedforward robust servo control problem presented in [14, 15]. Below, the above approach is described in a simplified manner which, however, serves our purpose well. Consider a linear time invariant system of the form

$$ \dot{x} = Ax + Bu + Dw, $$

$$ y = Cx + Fu + Hw, $$

(23)

where $x \in R^n$ is the state, $u \in R^m$ is the control input, $y \in R^p$ is the output, $w \in R^r$ are the disturbances, and A, B, C, D, F, and H are matrices of appropriate dimensions. Consider a control problem described as follows. Suppose the disturbances (t) are measurable. Given a desired output signal $y_{ref}(t)$, design a control algorithm that guarantees $y(t) \rightarrow yref(t)$ as $t \rightarrow +\infty$. This problem was addressed in [14, 15] in a very general setting. In this work, a simple case is addressed where both y_{ref} and (t) are assumed to be constant signals, $y_{ref}(t) \equiv y_{ref}$ and $w(t) \equiv w_m$. In this case, the following two conditions are necessary and sufficient for the existence of a linear time-invariant controller that solves the above described problem.

$$\text{rank} \left[B, AB, A^2 B, \ldots, A^{n-1} B \right] = n;$$

(24)

(ii)Consider

$$\text{rank} \begin{bmatrix} A & B \\ C & F \end{bmatrix} = n + p.$$

(25)

If the above two conditions hold (and only in this case), the linear time-invariant controller that solves the above described problem is given according to the formula

$$u = Kx + \mathcal{G}^\dagger y_{\text{ref}} + \mathcal{G}^* w_m,$$

(26)

where $K \in R^{n \times n}$ is the feedback gain matrix which is to be chosen such that $A - BK$ is stable and has the required dynamic properties

$$\mathcal{G} = -C(A - BK)^{-1} B,$$

(27)

$$\mathcal{G}^* = \mathcal{G}^\dagger C(A - BK)^{-1} D,$$

(28)

where $G^\dagger$ is the Moore-Penrose pseudoinverse of the matrix G in (27), defined by the formula

$$\mathcal{G}^\dagger = \mathcal{G}^T \left(\mathcal{G} \mathcal{G}^T \right)^{-1}.$$

(29)

The above described control approach can be applied to the problem of stabilization of the angular velocity of drilling as follows. Equations (22), which describe the rotational dynamics of a drilling system, can be rewritten in the form (23), where $:= [\omega_1 \, \phi \, \omega_2 \, I]^T \in R^4$, $u := V \in R^1$, $y := \omega_1 \in R^1$, $w := T \in R^1$, and the corresponding matrices are

$$A := \begin{bmatrix} \dfrac{-c_1}{J_1} & \dfrac{k}{J_1} & 0 & 0 \\ -1 & 0 & 1 & 0 \\ 0 & \dfrac{-k}{J_2} & \dfrac{-c_2}{J_2} & \dfrac{Kn}{J_2} \\ 0 & 0 & \dfrac{-Kn}{L} & \dfrac{-R}{L} \end{bmatrix}, \qquad B = \begin{bmatrix} 0 \\ 0 \\ 0 \\ \dfrac{1}{L} \end{bmatrix},$$

$$D = \begin{bmatrix} \dfrac{-1}{J_1} \\ 0 \\ 0 \\ 0 \end{bmatrix},$$

(30)

$$C = \begin{bmatrix} 1 & 0 & 0 & 0 \end{bmatrix}, \qquad F = \begin{bmatrix} 0 \end{bmatrix}, \qquad H = \begin{bmatrix} 0 \end{bmatrix}.$$

(31)

Below, we consider the drilling system with specific values of the parameters that are listed in Table1. With these values, the matrices A,B, and D become

$$A := \begin{bmatrix} -0.1123 & 1.2647 & 0 & 0 \\ -1 & 0 & 1 & 0 \\ 0 & -0.2231 & -0.2005 & 0.0204 \\ 0 & 0 & -8640 & -2 \end{bmatrix},$$

$$B = \begin{bmatrix} 0 \\ 0 \\ 0 \\ 200 \end{bmatrix}, \qquad D = \begin{bmatrix} -0.0027 \\ 0 \\ 0 \\ 0 \end{bmatrix},$$

(32)

while C, F, and H are given by (31).

Table 1: Numerical values for drilling system parameters

Parameter	Description	Value	Unit
J_1	BHA + drill string inertia	374	[kgm²]
J_2	Rotary table + drive inertia	2120	[kgm²]
c_1	BHA damping	42	[Nms/rad]
c_2	Rotary table damping	425	[Nms/rad]
k	Drill string stiffness	473	[Nms/rad]
R	Motor armature resistance	0.010	[Ω]
L	Motor armature inductance	0.005	[H]
K	Motor constant	6	[Vs]
n	Combined gear ratio for bevel and gear box	7.2	—
a	Drill bit radius	0.108	[m]
ζ	Ratio of drilling strength to drilling specific energy	0.7	—
M	Mass of drill string (28120 Kg) + BHA (25080 Kg)	53000	[kg]
$W_s - H_0$	Submerged weight W_s – applied weight from top of the Rig H_0	100 or 1000	[N]
K_f	Viscous friction coefficient	20	[Nm/rad]

For the above system, the necessary and sufficient conditions for stabilization (24), (25) are satisfied. Indeed, the stabilizability condition (24) is satisfied since

$$\text{rank} \left[B, AB, A^2B, \dots A^{n-1}B \right]$$

$$= \text{rank} \begin{bmatrix} 0 & 0 & 0 & 5.154273 \\ 0 & 0.000000 & 4.075472 & -8.968 \\ 0 & 4.075 & -8.968 & -700.339 \\ 200 & -400 & -34412 & 146307 \end{bmatrix} = 4.$$

(33)

On the other hand, the rank condition (25) is also satisfied because

$$\text{rank} \begin{bmatrix} A & B \\ C & F \end{bmatrix}$$

$$= \text{rank} \begin{bmatrix} -0.112299 & 1.264706 & 0 & 0 & 0 \\ -1 & 0 & 1 & 0 & 0 \\ 0 & -0.223113 & -0.200472 & 0.020377 & 0 \\ 0 & 0 & -8640 & -2 & 200 \\ 1 & 0 & 0 & 0 & 0 \end{bmatrix}$$

$$= 5.$$

(34)

Therefore, a controller of the form (26), (27), (28), and (29) guarantees that the angular velocity of the drill approach the reference angular velocity $\omega_1 \to \omega_{ref}$ as $t \to \infty$, while rejecting the disturbance T_b.

The design of controller (26), (27), (28), and (29) begins by choosing the desired location of the closed-loop system's poles. For the purpose of simulations presented below, we consider two specific set of poles. The first set, denoted by P_1, is chosen as follows::

$$P_1 := \begin{bmatrix} -10 & -2+2i & -2-2i & -4 \end{bmatrix}.$$

(35)

The set P_1 consists of two real poles and two complex conjugate poles. On the other hand, the set contains only poles on the real axis as follows:

$$P_2 = \begin{bmatrix} -5.5 & -2 & -4.5 & -1 \end{bmatrix}.$$

(36)

The feedback gain matrix K_1 such that the poles of $A - BK_1$ are located according to P_1 is

$$K_1 = \begin{bmatrix} 32.24 & 57.45 & -19.41 & 0.0784 \end{bmatrix}.$$

(37)

ThecoefficientsG^*,$G^\dagger$ in (26) are calculated according to the formulas (27)–(29); the results are

$$\mathscr{G}_1^* = 0.123497, \qquad \mathscr{G}_1^\dagger = 60.0844.$$

(38)

On the other hand, the feedbackmatrixK_2 such that the poles of $A - BK_2$ are located according to P_2 is

$$K_2 = \begin{bmatrix} -5.167 & 16.943 & -30.62 & 0.0534 \end{bmatrix}.$$

(39)

The corresponding coefficients $G_2^{\cdot}$, $G_2^{\dagger}$ are

$$\mathscr{G}_2^* = 0.037286, \qquad \mathscr{G}_2^\dagger = 9.603682.$$

(40)

Rock Stiffness Estimation

In the controller design presented above, it was assumed that the "hardness" of the rock, represented by the intrinsic specific energy ϵ, is constant and exactly known. This knowledge of ϵ was used explicitly in the controller design, in particular, in formula (21). In practical geological drilling, however, the hardness of different layers of rock lying underneath the surface can be different and usually is not exactly known beforehand. More specifically, different characteristics of the rock, such as hardness, density and porosity, typically remain constant through each layer, but differs from layer to layer. On the other hand, control engineers frequently deal with the problem of designing a controller without a priori knowledge of the exact values of one or more parameters involved in the process. Often, the processes can be robustly controlled without the actual knowledge of some of the parameters. In other cases, the unknown parameters can be identified using specially designed estimators. Below, a simple online estimator of the rock intrinsic specific energy ϵ is designed following the methods described in [16], and the resulting estimate is then used in the controller for for drilling system.

Specifically, during the cutting process, the torque-onbit T is produced by bit rock interaction, according to the formula

$$T = \frac{1}{2}a^2\epsilon d,$$

(41)

where a is the radius of drill bit, d is the depth of cut, and $\epsilon > 0$ is the intrinsic specific energy. The intrinsic specific energy $\epsilon > 0$ depends on the properties of the media and typically unknown beforehand. However, since the torque on bit (t) can typically be measured with advanced transducers located in the bottom hole assembly [17], $a > 0$ is constant and known, and (t) can be calculated according to formula (13), one can use the method described in the previous section to design an online estimation scheme for ϵ. In particular, considering $(1/2)a^2$ $d(t)$ as the input and torque-on-bit T as the measured output, one can follow the procedure described in the previous section to design an estimator for an unknown parameter ϵ.The predicted torqueon- bit $\hat{T}_b$ is defined according to the formula

$$\hat{T}(t) := \frac{1}{2}a^2\hat{\epsilon}d(t),$$

(42)

where $\epsilon(t)$ is the current estimate of actual rock strength ϵ. The algorithm for online estimation of the intrinsic specific energy ϵ has a form

$$\dot{\hat{\epsilon}} = \gamma_0\left(T - \hat{T}\right)\frac{1}{2}a^2d,$$

(43)

where $\gamma_0 > 0$ is an arbitrary gain.

A natural question regarding the algorithm (43) is if it guarantees the convergence of the parameter estimate to the true value of the parameter ϵ; mathematically, is$\hat{}(t) \rightarrow \epsilon$ as $t \rightarrow +\infty$. It is known [16], that the convergence can be guaranteed if the "input" signal $(1/2)^2$ (t) is persistently exciting. A signal (t) is said to be persistently exciting which is to say that there exist $\alpha_0 > 0$, $T_0 > 0$such that the inequality

$$\int_t^{t+T_0} u^2(\tau)\,d\tau \geq \alpha_0 T_0$$

(44)

holds for all t. In particular, (t) is persistently exciting if $u^2(t) \geq \alpha_0$ for all t. Since $d(t)$ is the depth of cut, we see that, during normal cutting process, $d(t) \geq d_0 > 0$, which results in persistent excitation of the input $(1/2)a^2\,d(t)$. The parameter convergence $\hat{}(t) \to \epsilon$, therefore, is guaranteed during normal cutting process. This is also confirmed by the simulation results presented below.

The obtained estimate of the rock strength $\hat{}\epsilon$ is then used in the control algorithm. Specifically, in the original formulation of the control algorithm, for a given reference vertical velocity V_{ref}, the reference rotational velocity ω_{ref} is calculated according to formula (21), which depends on the parameter ϵ. In case ϵ is unknown, it is substituted by its estimate $\hat{}(t)$ obtained above. The new formula for ω_{ref} has the form

$$\omega_{ref} := \frac{2\pi a \zeta \hat{\epsilon}}{((W_s - H_0)/v_{ref}) - K_f}.$$

(45)

The obtained estimate of the rock stiffness ϵ will also be used to update the stiffness of the virtual spring in the haptic teleoperator drilling system described below.

Simulation Results

In this Section, some results of simulations of the drilling control system with intrinsic specific energy estimator are presented. The vertical motion of the drilling system is described by (15), and it is interconnected with the rotational dynamics (5) through nonlinear equation (13) that describes the depth of cut (t). For a given reference velocity of the vertical penetration $V_{ref} > 0$, the corresponding reference rotational velocity ω_{ref} is calculated according to formula (45). The controller (21), (26)–(29) has been implemented to guarantee that the angular velocity of the drill bits $\omega_1(t)$ tracks ωref, which in turn stabilizes the vertical penetration velocity $V(t)$ converges to V_{ref}. The algorithm

(43) provides an estimate of the intrinsic specific energy parameter $\hat{\epsilon}$ which is then used in the calculation of the reference angular velocity according to formula (45). Specific values of the parameters appearing in these equations are given in Table 1. The simulations are carried out using MATLAB, where the integration step for each simulation is. equal to 0.005 s. The feedback gain matrix is chosen $K = K_1$, where K_1 is defined by (37).

In the simulations described below, the performance of the system was evaluated for different values of actual intrinsic specific energy ϵ, different gains γ_0 and different values of the applied weight $W_0 :=$ $W_s - H_0$. Figures 5 and 6 show the response of the vertical penetration velocity $V(t)$, the intrinsic specific energy estimate $\hat{\epsilon}(t)$, the torque-on-bit $T(t)$, the predicted value of the torque-on-bit $\hat{T}(t)$, and the rotational velocity $\omega_1(t)$, all for the case where the applied weight on bit $W_0 =$ 5000N, the intrinsic specific energy $\epsilon = 20$MPa, and the desired vertical velocity V_{ref} is set to 0.005m/s. The estimator gain is set to $\gamma_0 = 5 \cdot 10^9$. The plots show that $V(t)$ converges to V_{ref} in less than 8 sec whereas the estimate $\hat{\epsilon}(t)$ converges to the actual value of ϵ in less than 4 sec. Figures 7 and 8 show the output responses of described parameters where W_0 = 2500N and the desired vertical velocity V_{ref} is set to 0.01m/s. It can be clearly seen that the convergence becomes slower with reducing the applied weight on the drill string W_0; specifically, both $V(t)$ and $\hat{\epsilon}(t)$ approach their reference values in about 12 sec. Reducing W_0 also results in that ω_{1ref} increases, the steady-state value of $T_b(t)$ drops to around 200N, and the steady-state value of $d(t)$ also drops to less than 2 mm. On the other hand, Figures 9 and 10 demonstrate the response of the system with the same parameters except the intrinsic specific energy ϵ is reduced to 5MPa. This results in decreased convergence time for $V(t)$ and $\hat{\epsilon}(t)$. The steady state value of rotational velocity $\omega_1(t)$ is also decreased to under 10 rad/s, and steady state value of the depth of cut $d(t)$ is increased to 6.5 mm. Figures 11 and 12 present the output response for the case where the estimator gain is decreased to $\gamma_0 = 5 \cdot 10^8$, while the rest of the parameters are the same as in the last simulation except the intrinsic specific energy is set to $\epsilon = 10$MPa. The resulting response is predictably characterized by much slower convergence, which takes about 25 sec for $V(t)$ and $\hat{\epsilon}(t)$ to approach their steady-state values. Finally, Figures 13 and 14 correspond to to the case where $W_0 = 5000$ N, $\epsilon = 20$MPa, and $\gamma_0 = 1 \cdot 10^8$.

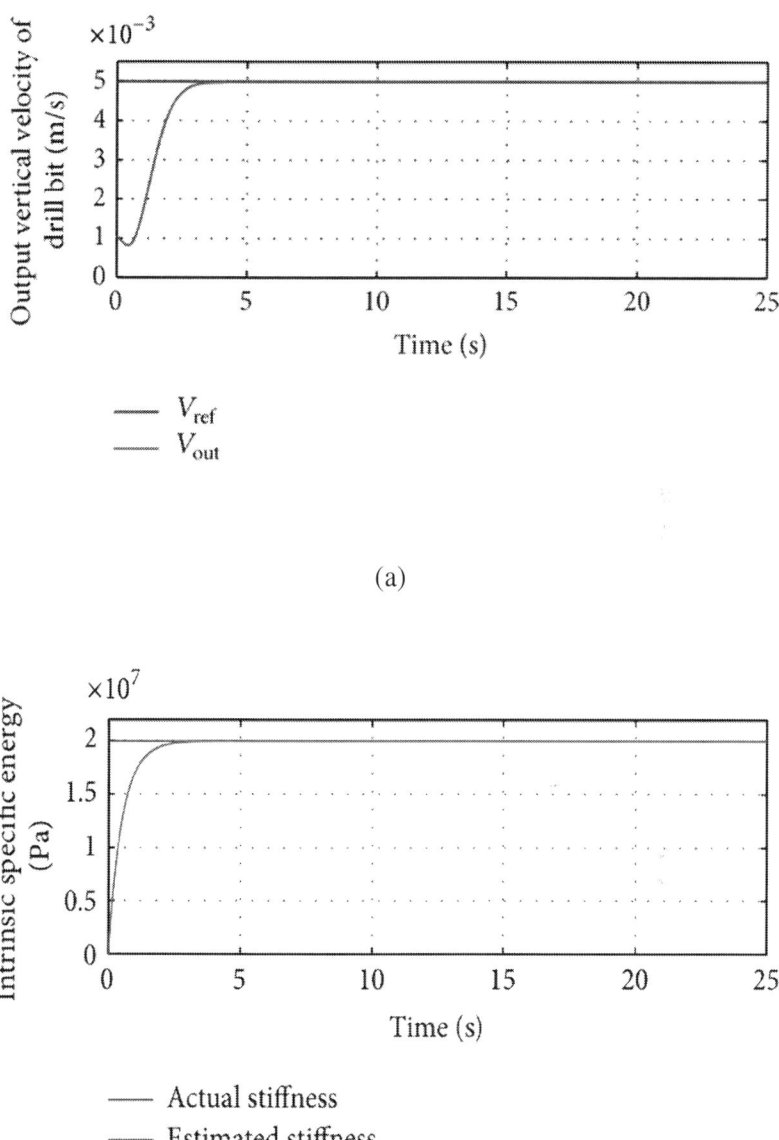

(a)

(b)

Figure 5: Response of the vertical velocity $V(t)$ (a) and the intrinsic specific energy estimate $\epsilon(t)$ (b) for $W_0 = 5000$ N, $\epsilon = 20$MPa, and $\gamma_0 = 5 \cdot 10^9$.

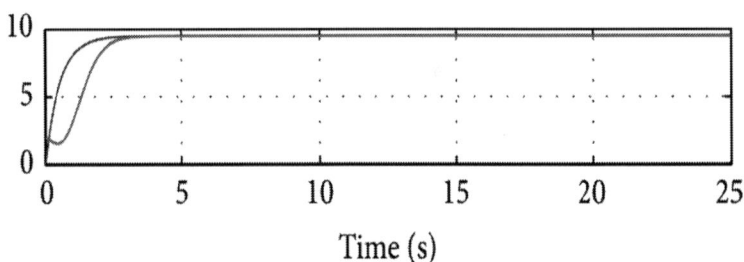

(a)

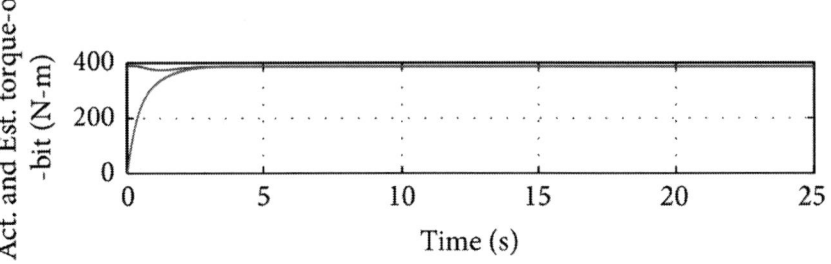

(b)

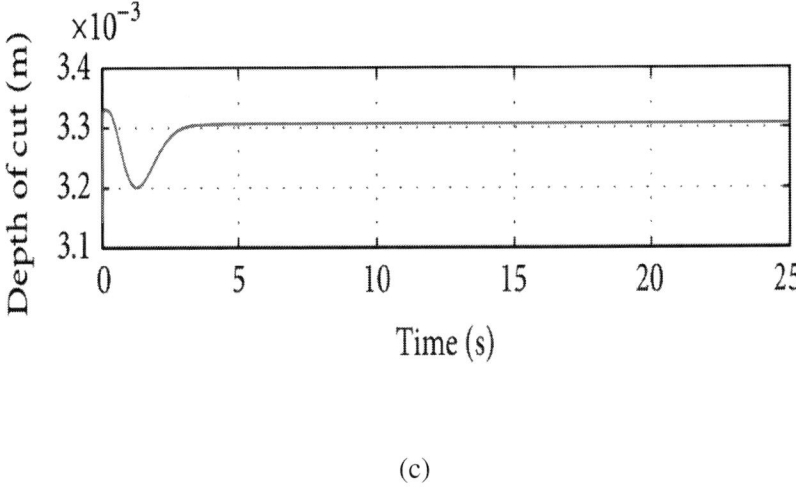

(c)

Figure 6: Response of rotational velocity $\omega_1(t)$ (a), torque-on-bit $T(t)$ vresus estimated torque-on-bit $\hat{T}(t)$ (b), and the depth of cut $d(t)$ (c) for$W0 = 5000$ N, $\epsilon = 20$MPa, and $\gamma_0 = 5 \cdot 10^9$.

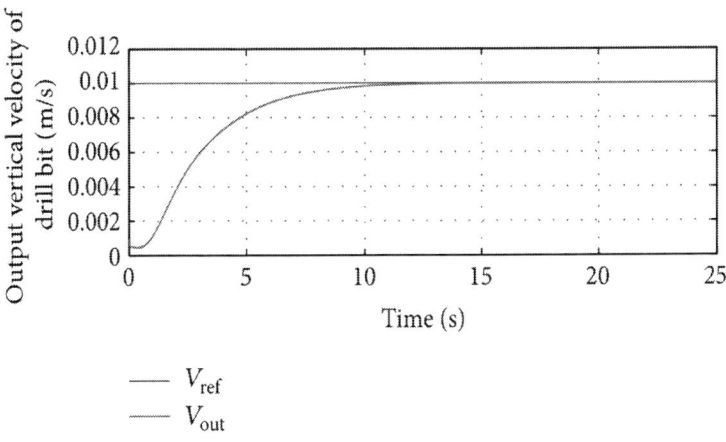

(a)

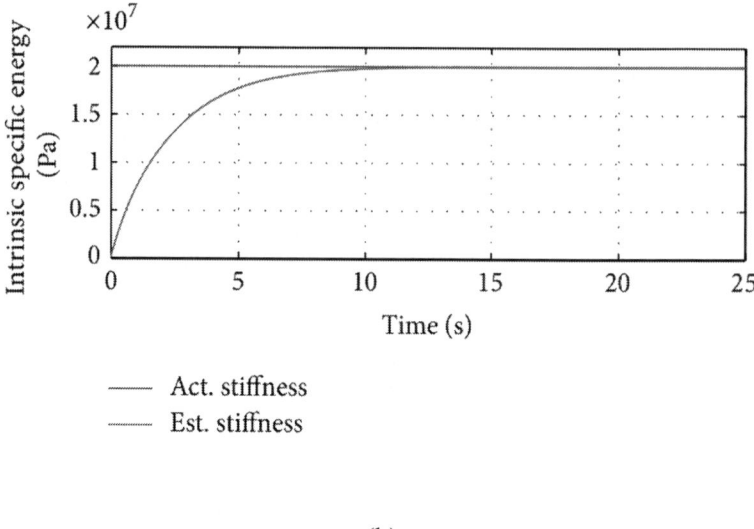

(b)

Figure 7: Response of the vertical velocity V(t) (a) and the intrinsic specific energy estimate $\hat{\epsilon}(t)$ (b) for $W_0 = 2500$ N, $\epsilon = 20$MPa, and $\gamma_0 = 5 \cdot 10^9$.

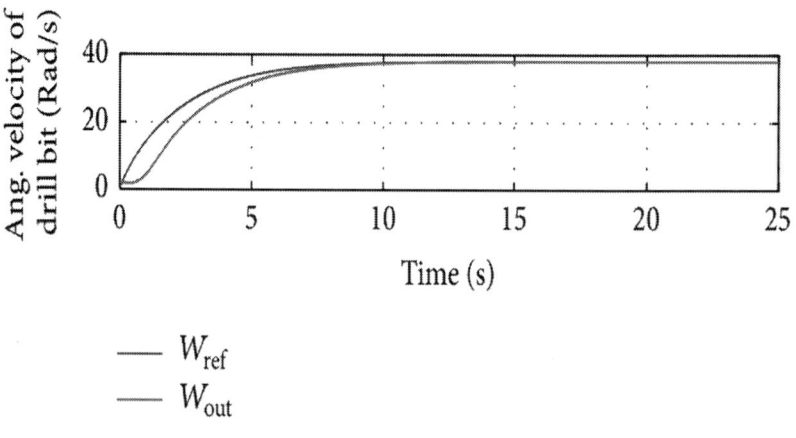

(a)

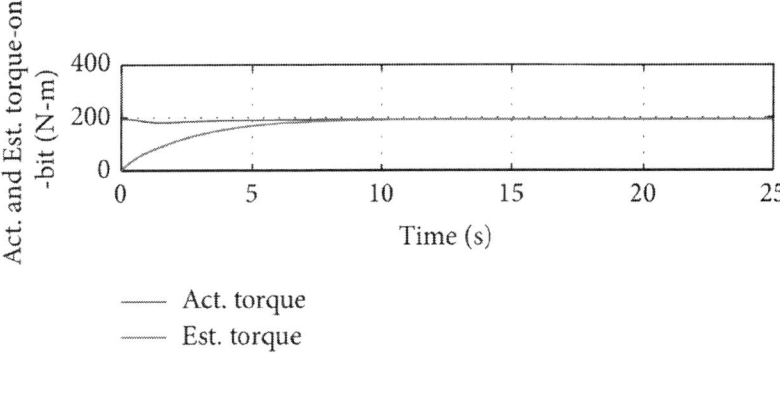

(b)

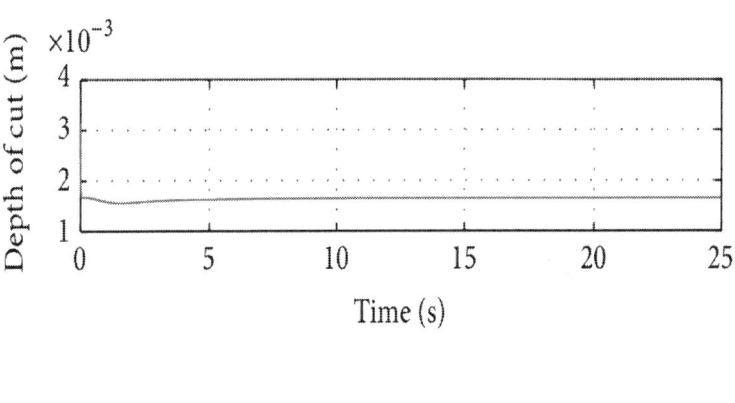

(c)

Figure 8: Response of rotational velocity $\omega 1(t)$ (a), torque-on-bit $T(t)$ versus estimated torque-on-bit $T(t)$ (b), and the depth of cut $d(t)$ (c) forW_0 = 2500 N, ϵ = 20MPa, and $\gamma 0 = 5 \cdot 10^9$.

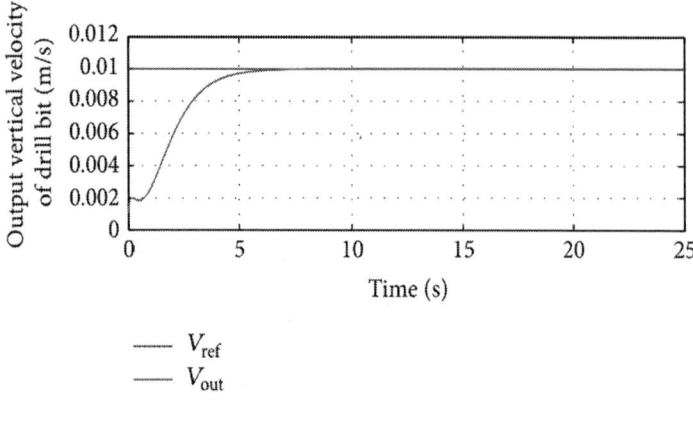

(a)

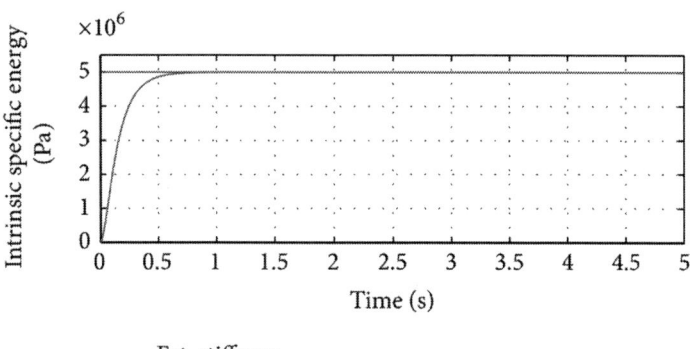

(b)

Figure 9: Response of the vertical velocity V(t) (a) and the intrinsic specific energy estimate $\epsilon(t)$ (b) for $W_0 = 2500$ N, $\epsilon = 5$MPa, and $\gamma_0 = 5 \cdot 10^9$.

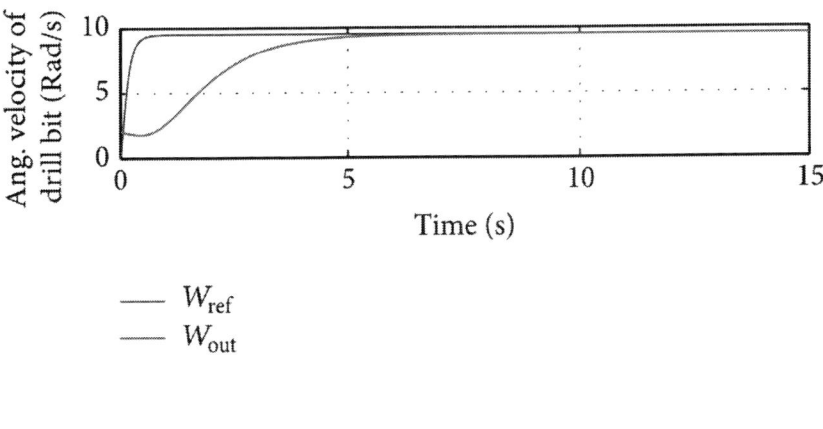

(a)

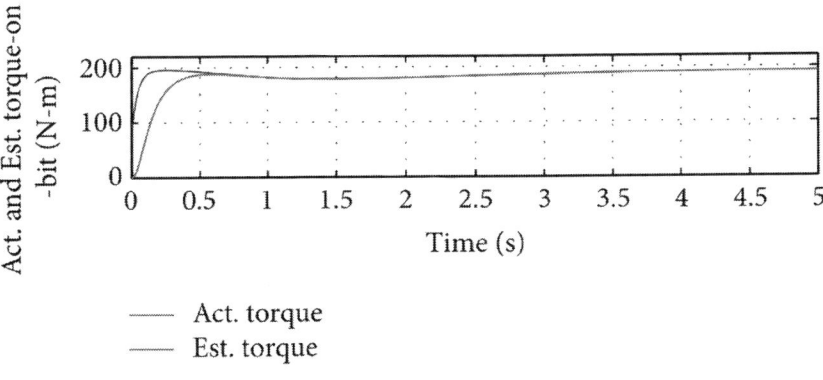

(b)

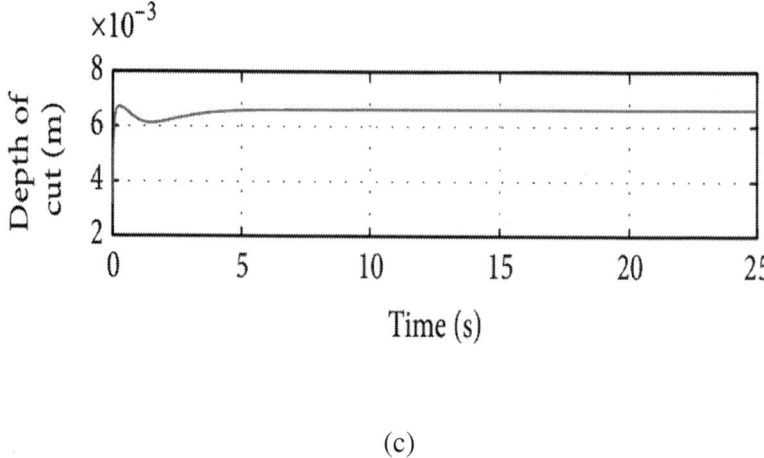

(c)

Figure 10: Response of rotational velocity $\omega_1(t)$ (a), torque-on-bit $T(t)$ versus estimated torque-on-bit $\hat{T}(t)$ (b), and the depth of cut $d(t)$ (c) for $W_0 = 2500$ N, $\epsilon = 5$MPa, and $\gamma_0 = 5 \cdot 10^9$.

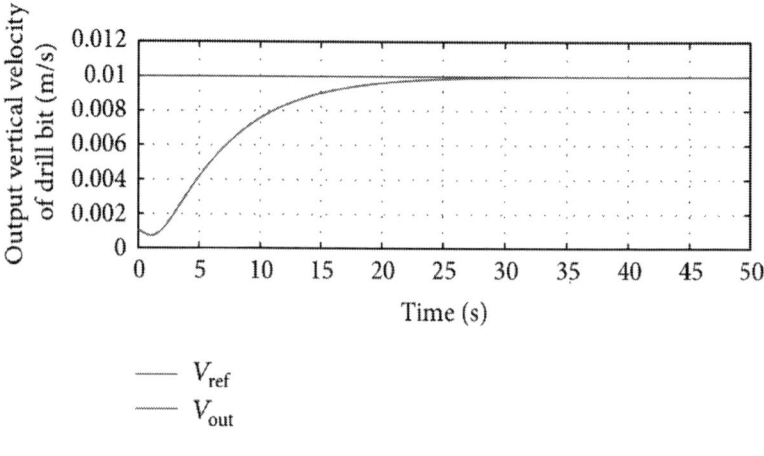

(a)

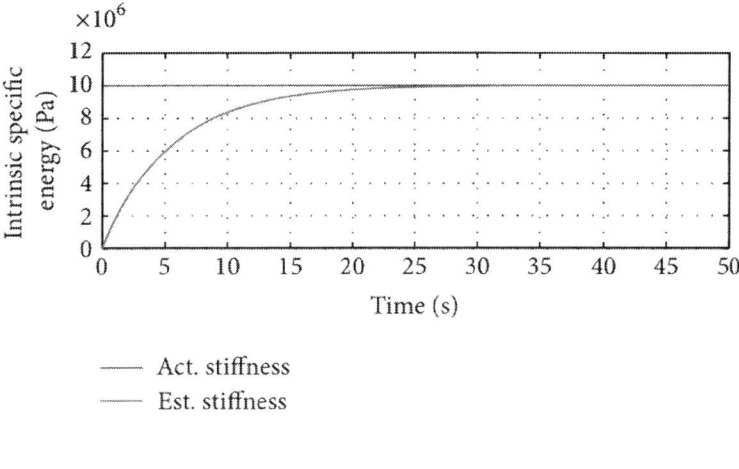

(b)

Figure 11: Response of the vertical velocity V(t) (a) and the intrinsic specific energy estimate $\epsilon(t)$ (b) for W_0 = 2500 N, ϵ = 10MPa, and γ_0 = 5 · 10^8.

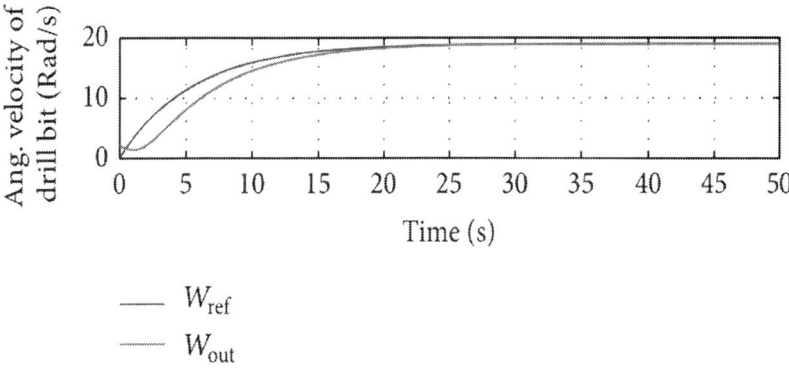

(a)

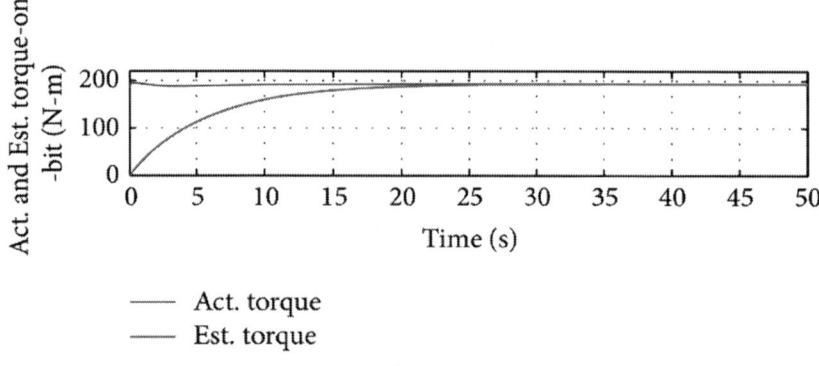

(b)

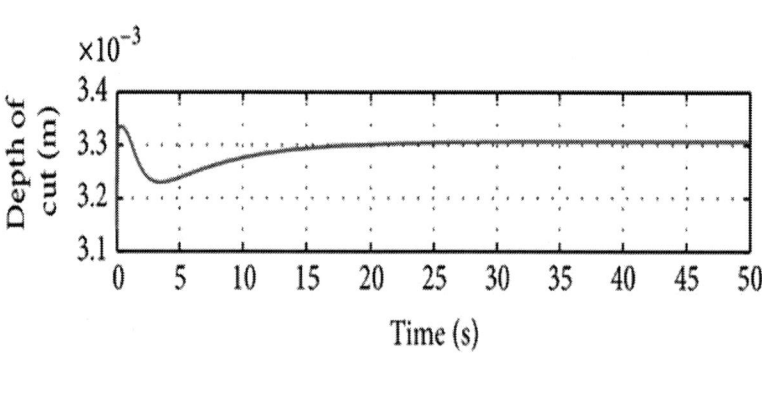

(c)

Figure 12: Response of rotational velocity $\omega_1(t)$ (a), torque-on-bit $T(t)$ versus estimated torque-on-bit $T(t)$ (b), and the depth of cut $d(t)$ (c) for $W_0 = 2500$ N, $\epsilon = 10\text{MPa}$, and $\gamma_0 = 5 \cdot 10^8$.

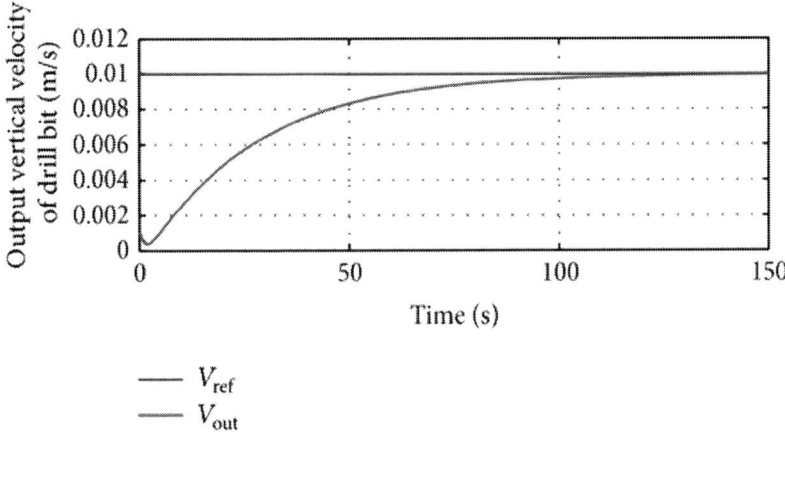

(a)

(b)

Figure 13: Response of the vertical velocity V(t) (a) and the intrinsic specific energy estimate $\epsilon(t)$ (b) for $W_0 = 2500$ N, $\epsilon = 20$MPa, and $\gamma_0 = 10^8$.

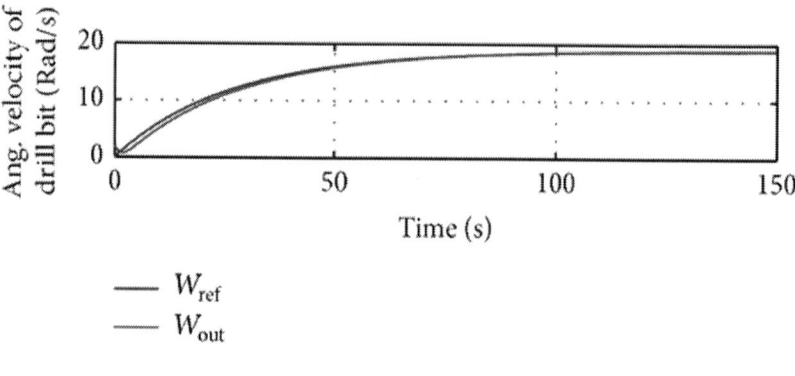

(a)

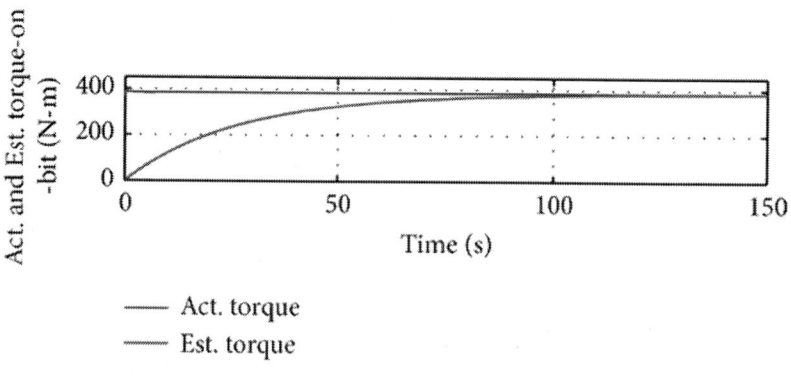

(b)

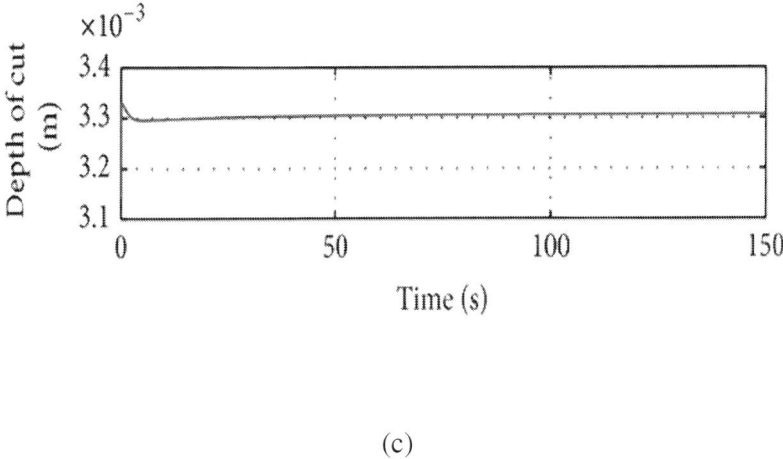

(c)

Figure 14: Response of rotational velocity $\omega_1(t)$ (a), torque-on-bit $T(t)$ versus estimated torque-on-bit $\hat{T}(t)$ (b), and the depth of cut $d(t)$ (c) for $W0 = 2500$ N, $\epsilon = 20$MPa, and $\gamma_0 = 10^8$.

Overall, simulation results show that the control system with intrinsic specific energy estimation demonstrate good stability and performance characteristics for a wide range of the parameters. In particular, the vertical velocity converges to the desired value, and the estimate of the intrinsic specific energy (t) converges to an actual value of ϵ.

TELEROBOTIC DRILLING SYSTEM WITH HAPTIC FEEDBACK

In this section, a telerobotic drilling system with haptic feedback is designed and experimentally evaluated. Haptics can be defined as the physical or virtual interaction through touch sensation for the purpose of perception and manipulation of objects [18, 19]. Haptic feedback provides the operator with kinaesthetic clues of the physical features of virtual or real remote environment. The structure of a telerobotic drilling system with haptic feedback is shown in Figure 15. In this system, the human operator controls the drilling process using a haptic device. Specifically, the position of an end-effector of the haptic device defines the reference vertical velocity of the drilling. The reference vertical velocity is then transmitted to the drilling control system, designed in above in Section 3, which stabilizes the actual vertical penetration velocity to the level equal to the reference vertical velocity. On the other hand, an estimate of the intrinsic specific energy , which is generated online by an estimator described in above in Section 3.3, is sent back to the haptic device. The end-effector of the haptic device interacts with a virtual spring of variable stiffness; the stiffness of this virtual spring is updated in real time proportionally to the current estimate of the intrinsic specific energy (t). Thus, the telerobotic drilling system provides haptic feedback to the human operator which creates an intuitive feeling of the hardness of the remotely drilled material.

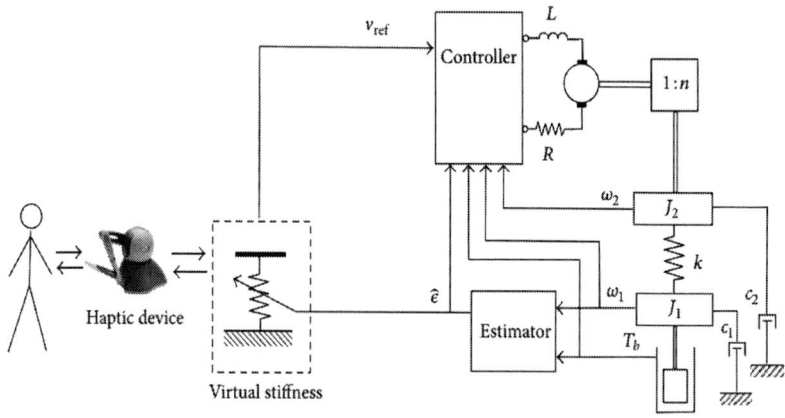

Figure 15: The structure of a telerobotic drilling system.

Experimental Setup

The above described telerobotic drilling system is implemented in a semiexperimental setup, as follows. The setup consists of a PC based on Intel Pentium 4 processor with operational frequency of 1 GHz and RAM of 1 GB, and a PHANTOM Omni Haptic device, a product from SensAble Technologies Inc. The PHANTOM Omni Haptic device is designed for kinematic interaction with the virtual or real environment while providing the kinesthetic feedback to the operator. The device is equipped with a pen-based stylus, and is able to provide three degrees-of-freedom force feedback. The human operator uses the haptic device to (i) generate a desired vertical velocity $v_{ref}(t)$ which is used as an input to the drilling control system, and (ii) to haptically perceive the stiffness ϵ of the rock layers. The remaining parts of the above described telerobotic system, including the drill string and drive system, the drilling process, as well as the control and estimation algorithms, are simulated in real time in virtual environment which is implemented using the Open Haptics tool kit and Microsoft Visual C++.

The human operator assigns the desired velocity $v_{ref}(t)$ by controlling the position of the end-effector of the PHANTOM device along its vertical (Y) axis. More exactly, a specific range along the y-axis is assigned to each desired vertical velocity, as follows:

$v_{ref}(t) = 0.001$ m/s if the position of stylus $y_n(t)$ is ≥ 80 mm;

$v_{ref}(t) = 0.003$ m/s if the position of stylus $y_n(t)$ is ≤ 80 mm and ≥ 60 mm;

$v_{ref}(t) = 0.005$ m/s if the position of stylus $y_n(t)$ is ≤ 60 mm and ≥ 40 mm;

$v_{ref}(t) = 0.008$ m/s if the position of stylus $y_n(t)$ is ≤ 40 mm and ≥ 25 mm;

$v_{ref}(t) = 0.01$ m/s if the position of stylus $y_n(t)$ is ≤ 25 mm and ≥ 10 mm;

$v_{ref}(t) = 0.015$ m/s if the position of stylus $y_n(t)$ is ≤ 10 mm and ≥ 0 mm;

Another function of the haptic device is to allow the human operator to feel the stiffness of the rocks. As explained above, this is achieved by updating the stiffness of the virtual spring proportionally to the current estimate of the rock stiffness (intrinsic specific energy $\hat{e}(t)$). The coefficient of proportionality between the estimate of the intrinsic specific energy (with units of Pascals) and the stiffness of the virtual spring (with units of is N/m) is set in our experiments equal to 10^{-7}. The feedback force $F_{est}(t)$ due to the virtual spring is therefore calculated according to the formula

$$F_{est}(t) = 10^{-7} \cdot \hat{e}(t) \cdot y_n(t).$$

(46)

Experimental Results

In this Section, some experimental results are presented. In these experiments, we have attempted to simulate a real drilling case scenario where the composition, characteristics, and types (which all contribute to the intrinsic specific energy) of various rock strata vary at different depths during drilling. Specifically, in every experiment, several layers of rocks with different intrinsic specific energy ϵ ranging from 4MPa to 60MPa are simulated. In all experiments presented here, the applied weight$W0 = 5000$ N; the rest of the parameters, if not explicitly mentioned, are same as in Section 3.4.

In the experiment shown in Figures 16, 17, and 18, three layers of rocks with different intrinsic specific energy ϵ are simulated. The top layer has the stiffness of 5MPa and its thickness is 20 cm from the surface. The second layer has a stiffness value of 12MPa and lies between 20cm and 30cm from the surface (total thickness is 10 cm). The third layer starts at the depth of 30 cm and continues downward. It has a stiffness value of 20MPa. The experiment is performed with the estimator gain $\gamma_0 = 10^9$. Figure 16 shows the actual intrinsic specific energy (t) and its estimate (t) on the top graph, and the reflected force $F_{est}(t)$ on the bottom graph. Due to high estimator gain, (t) quickly tracks $\epsilon(t)$ for all three layers as the drill bit progressed cutting through these layers. Figure 17 shows the vertical velocity $V_{out}(t)$ and the reference vertical velocity $V_{ref}(t)$ on the top graph, and the reference rotational velocity $\omega_1 d(t)$ and the actual drill bit rotational velocity $\omega1(t)$ at the bottom graph. Finally, Figure 18 shows the behaviour of the actual torque-on-bit (t) and the estimated torque (t), along with depth of cut (t). These plots show that the system is stable and demonstrates good performance; in particular, all the output variables track their desired (reference) trajectories.

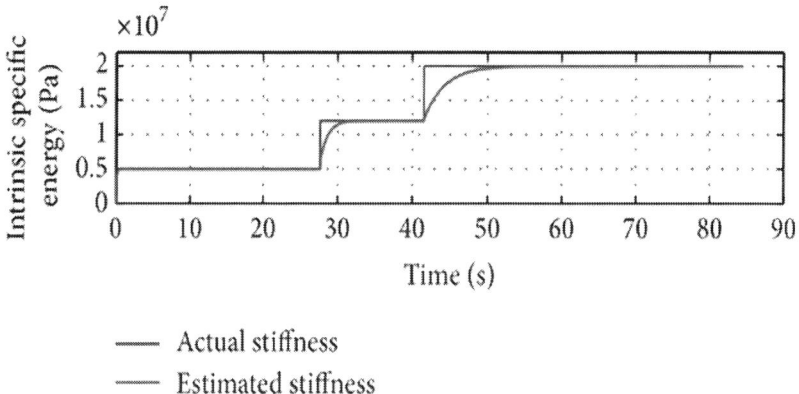

(a)

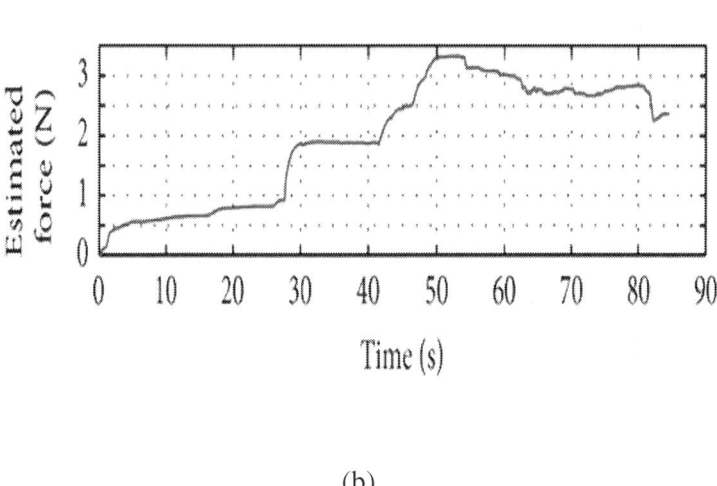

(b)

Figure 16: Experiment 1: Actual stiffness (t) versus the estimated stiffness (t) (a); the reflected force $F_{est}(t)$ (b).

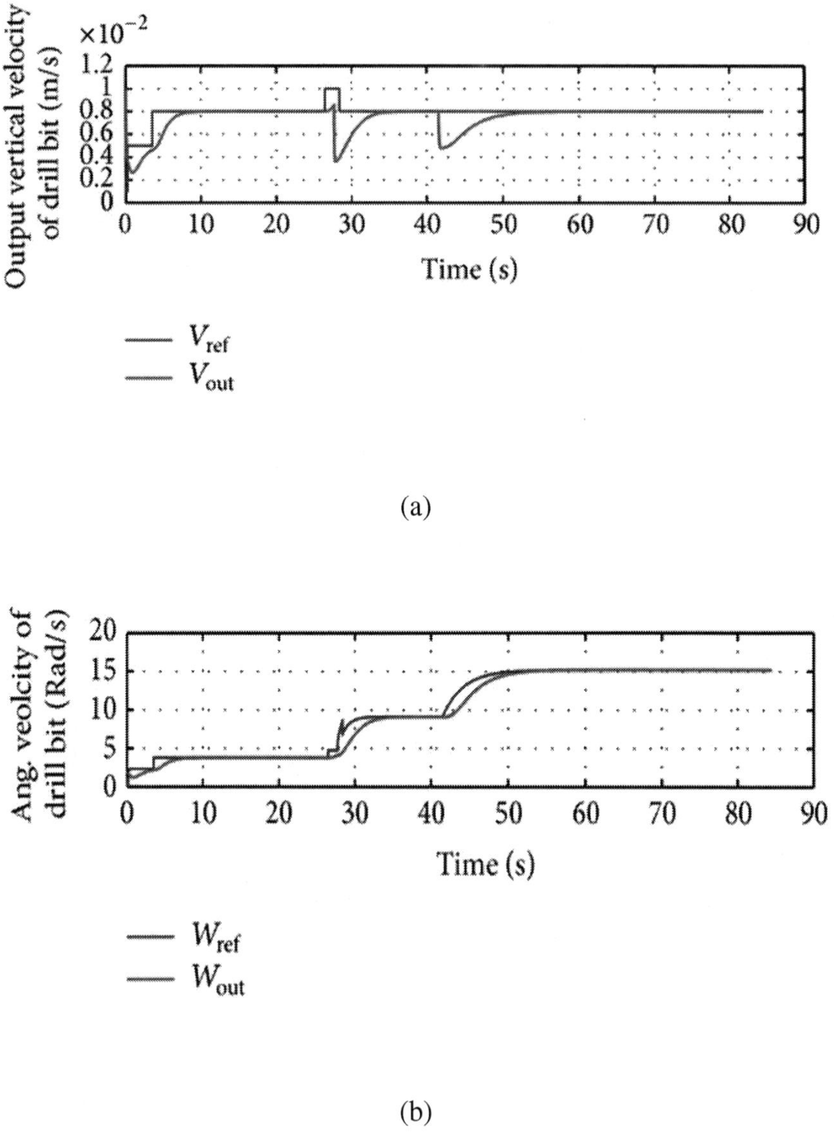

(a)

(b)

Figure 17: Experiment 1: Output vertical velocity Vout(t) versus reference vertical velocity Vref(t) (a); output rotational velocity of the drill bit $\omega_1(t)$ versus reference rotational velocity $\omega_1 d(t)$ (b).

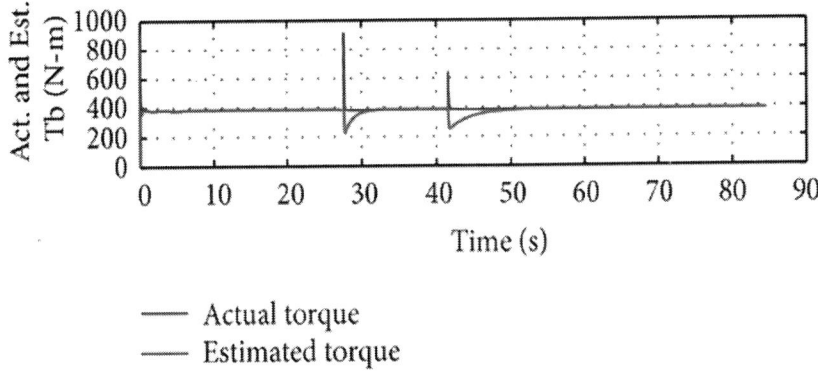

(a)

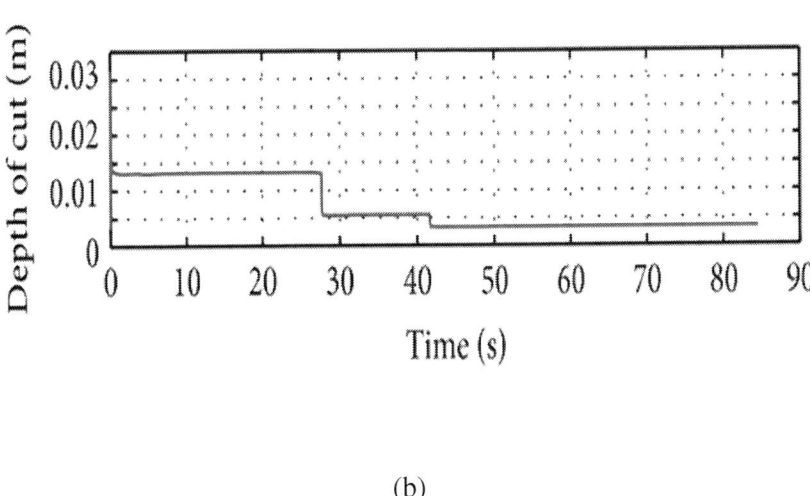

(b)

Figure 18: Experiment 1: Torque-on-bit (t) versus estimated torque-on-bit (t) (a); depth of cut $d_{cut}(t)$ (b)..

Another set of experimental results is presented in Figures 19–21, where the estimator gain is increased to $\gamma_0 = 5 \cdot 10^9$, and the depth of the rock layers and their corresponding stiffness values have been altered. Specifically, the first rock layer has depth 20 cm and the intrinsic specific energy ϵ is set to 20MPa for this layer. Second layer lies between 20 cm and 40 cm with ϵ = 40MPa. The third layer lies below 40 cm, and its intrinsic specific energy ϵ = 60MPa. Figure 19 shows the corresponding plots of (t), (t) and $F_{est}(t)$. Figure 20 shows the response of $V_{ref}(t)$ and $V_{out}(t)$ on the top graph, and the responses of $\omega 1 d(t)$ and $\omega_1(t)$ on the bottom graph, respectively. The response of (t) and (t) along with $d(t)$ are shown in Figure 21.Overall, our experiments demonstrate stability and good performance of the designed telerobotic drilling system with haptic feedback, for a wide range of parameters and control gains.

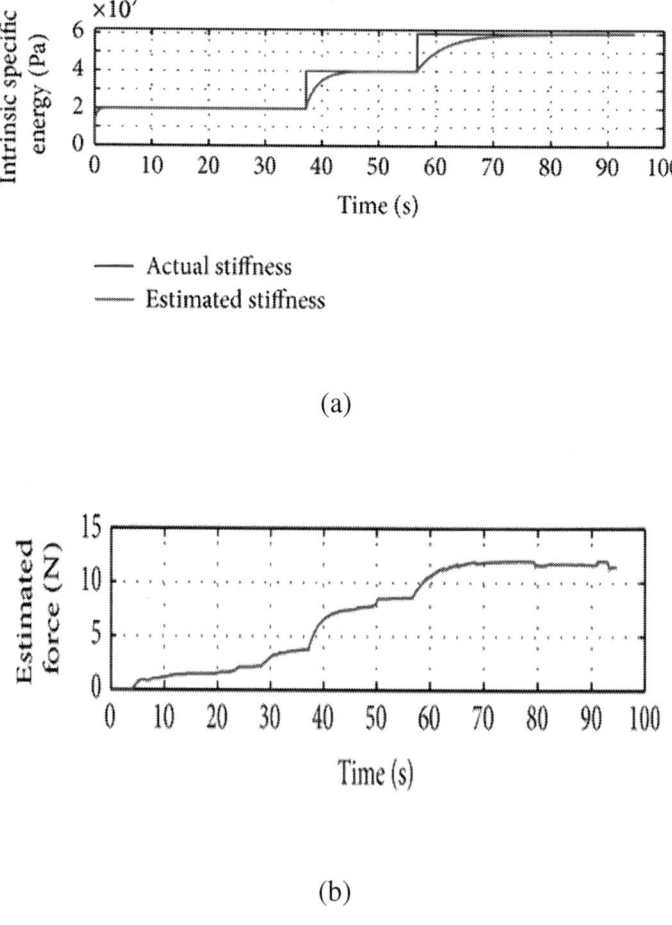

(a)

(b)

Figure 19: Experiment 2: Actual stiffness (t) versus estimated stiffness (t) (a); reflected force $F_{est}(t)$ (b).

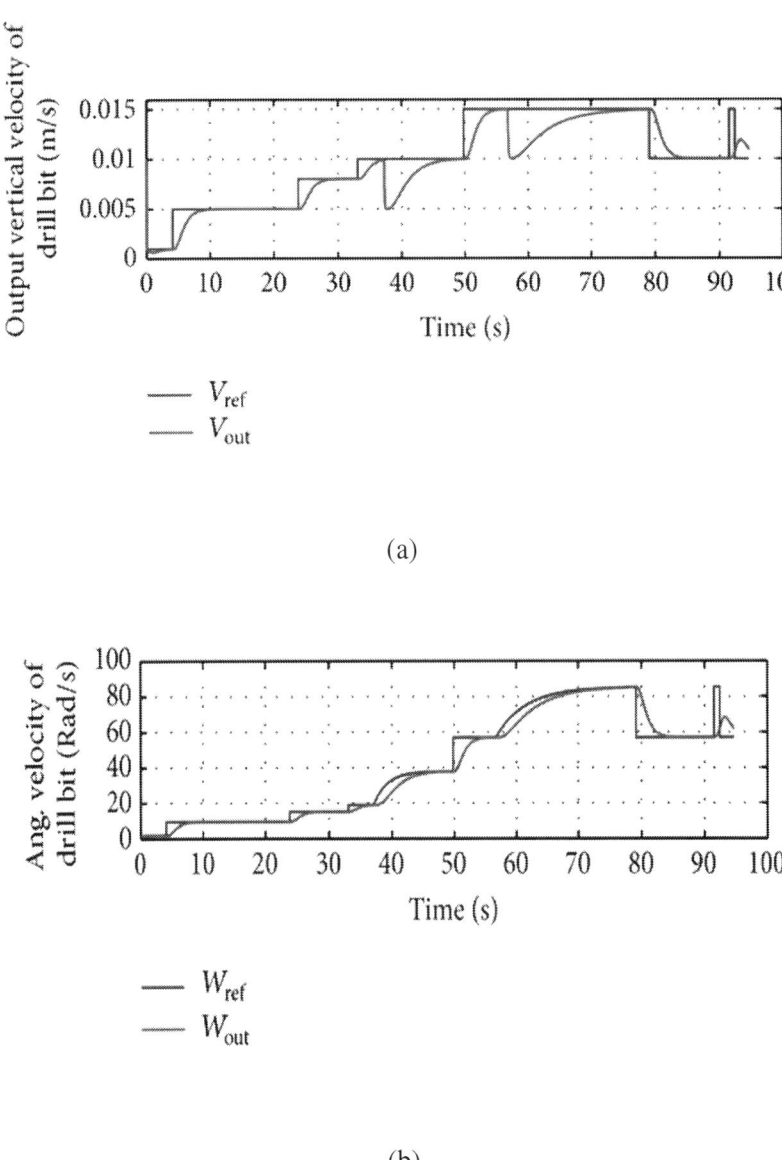

(a)

(b)

Figure 20: Experiment 2: Output vertical velocity Vout(t) versus reference vertical velocity $V_{ref}(t)$ (a); output rotational velocity of the drill bit $\omega_1(t)$ versus reference rotational velocity $\omega_1 d(t)$ (b).

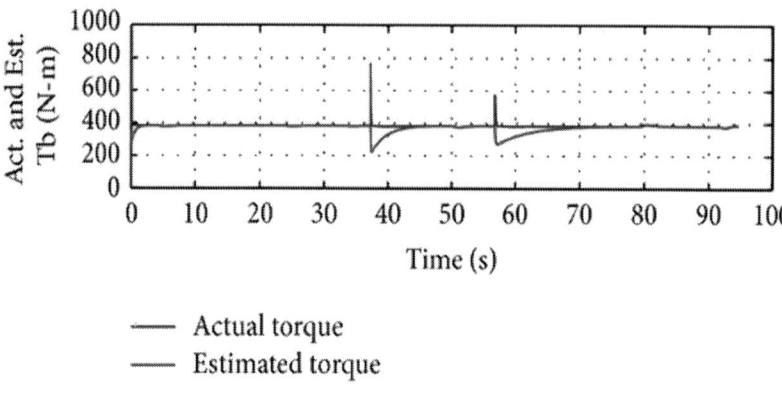

(a)

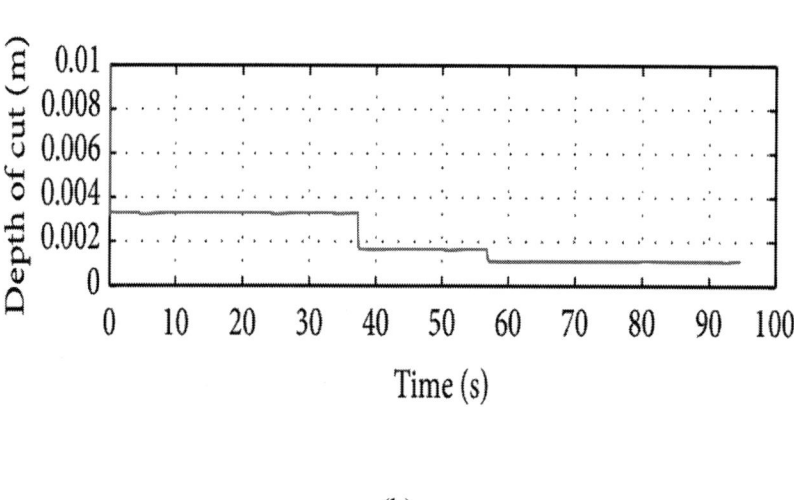

(b)

Figure 21: Experiment 2: Torque-on-bit (t) versus estimated torque-on-bit (t) (a); depth of cut $d_{cut}(t)$ (b).

CONCLUSIONS

This paper deals with control design for a teleoperator system with haptic feedback for an oil well drilling process. A mathematical model of the drilling process was described, and the control algorithm was designed that guarantees the convergence of the vertical penetration velocity to an arbitrary reference value. The control algorithm has a cascaded structure, where the velocity of vertical penetration is controlled indirectly through stabilization of the rotational motion of the drill bit. In order to guarantee the convergence of the angular velocity to a desired value in the presence of disturbances in the form of torque-on-bit, a robust servo controller was designed. However, the design of such controller depends on the parameter of environment called the intrinsic specific energy, which is generally unknown beforehand. To solve this issue, an online parameter estimator was designed that provides an estimate of the intrinsic specific energy. This estimate is substituted for the actual value of the parameter in the control algorithm, and the corresponding adaptive control system is evaluated through simulations. Finally, a telerobotic drilling system with haptic feedback is designed and verified through semi-experiments. The haptic feedback for the human operator is provided by creating a virtual spring that interacts with the haptic device; the stiffness of the spring is adjusted in real time depending on the current estimate of the intrinsic specific energy. Semi-experiments are conducted using PHANTOM Omni Haptic device, where the drilling process model is implemented in C++ environment, and the haptic feedback is provided to the human operator.

There exists a number of challenges associated with the real-life drilling operation that were not addressed in our paper. In particular, the frictional forces at the contact were neglected in our analysis, while in reality they may play significant role in the drilling process. Also, in real-life drilling systems, the rotational velocity and the penetration rate are typically measured at the surface while the torque-on-bit should ideally be measured above the bit; thus, there exists a problem of synchronizing the data obtained at the surface with those obtained at the bit. The issue of communication delay between the haptic device and the drilling process is also not addressed. These, as well as more detailed experimental evaluation of the designed system, are the topics for future research.

ACKNOWLEDGMENTS

This work was supported by the Natural Sciences and Engineering Research Council (NSERC) of Canada under Discovery Grant RGPIN1510.

REFERENCES

1.   P. Corke, J. Roberts, J. Cunningham, and D. Hainsworth, "Mining robots," in Springer Hand-Book of Robotics, pp. 1127–1150, Springer, New York, NY, USA, 1st edition, 2007.

2.   E. Jackson and D. Clarke, "Subsea excavation of seafloor massive suiphides," in Proceedings of the IEEE Oceans Conference, Vancouver, Canada, October 2007.

3.   N. Ridley, S. Graham, and S. Kapusniak, "Sea oor production tools for the resources of the future," in Proceedings of the Offshore Technology Conference, Houston, Tex, USA, May 2011.

4.   M. N. Wendt and G. A. Einicke, "Development of a water-hydraulic self-propelled robotic drill for underground mining," in Field and Service Robotics, vol. 25, pp. 355–366, Springer, New York, NY, USA, 2006.

5.   G. Baiden, "Telerobotic lunar habitat construction and mining: a miner's perspective," inProceedings of the 9th International Symposium on Artificial Intelligence, Robotics and Automation for Space (i-SAIRAS '08), Los Angeles, Calif, USA, 2008.

6.   J. Lee, I. Hwang, K. Kim, S. Choi, W. K. Chung, and Y. S. Kim, "Cooperative robotic assistant with drill-by-wire end-effector for spinal fusion surgery," Industrial Robot, vol. 36, no. 1, pp. 60–72, 2009.

7.   B. Glass, H. Cannon, S. Hanagud, and J. Frank, "Drilling automation for subsurface planetary exploration," in Proceedings of the 8th International Symposium on Artificial Intelligence, Robotics and Automation in Space (i-SAIRAS '05), pp. 205–209, Munich, Germany, September 2005.

8.   F. Poletto and F. Miranda, Seismic While Drilling: Fundamentals of Drill-Bit Seismic For Exploration, vol. 35 of Handbook of

Geophysical Exploration: Seismic Exploration, Elsevier, San Diego, Calif, USA, 2004.

9. J. D. Jansen and L. van den Steen, "Active damping of self-excited torsional vibrations in oil well drillstrings," Journal of Sound and Vibration, vol. 179, no. 4, pp. 647–668, 1995.

10. T. Richard, C. Germay, and E. Detournay, "Self-excited stick-slip oscillations of drill bits,"Comptes Rendus, vol. 332, no. 8, pp. 619–626, 2004.

11. T. Richard, C. Germay, and E. Detournay, "A simplified model to explore the root cause of stick-slip vibrations in drilling systems with drag bits," Journal of Sound and Vibration, vol. 305, no. 3, pp. 432–456, 2007.

12. M. Zamanian, S. E. Khadem, and M. R. Ghazavi, "Stick-slip oscillations of drag bits by considering damping of drilling mud and active damping system," Journal of Petroleum Science and Engineering, vol. 59, no. 3-4, pp. 289–299, 2007.

13. E. Detournay, T. Richard, and M. Shepherd, "Drilling response of drag bits: theory and experiment," International Journal of Rock Mechanics and Mining Sciences, vol. 45, no. 8, pp. 1347–1360, 2008.

14. E. J. Davison, "The feedforward control of linear multivariable time-invariant systems,"Automatica, vol. 9, no. 5, pp. 561–573, 1973.

15. E. J. Davison, "Multivariable tuning regulators: the feedforward and robust control of a general servomechanism problem," IEEE Transactions on Automatic Control, vol. 21, no. 1, pp. 35–47, 1976.

16. P. A. Ioannou and J. Sun, Robust Adaptive Control, Prentice Hall, New York, NY, USA, 1996.

17. B. P. Peltier, "Drilling monitor with downhole torque and axial load transducers," US Patent 4 695 957, September 1987.

18. T. H. Massie and J. K. Salisbury, "The PHANTOM haptic interface: a device for probing virtual objects," in Proceedings of the ASME Winter Annual Meeting, Symposium on Haptic Interfaces for Virtual Environment and Teleoperator Systems, pp. 295–299, Chicago, Ill, USA, November 1994.

19. K. Salisbury, F. Conti, and F. Barbagli, "Haptic rendering: introductory concepts," IEEE Computer Graphics and Applications, vol. 24, no. 2, pp. 24–32, 2004.

Chapter 2

Design and Development of Turbodrill Blade Used in Crystallized Section

Wang Yu, Yao Jianyi and Li Zhijun

Key Laboratory on Deep Geo-Drilling Technology of the Ministry of Land and Resources, China University of Geosciences, Beijing 100083, China

ABSTRACT

Turbodrill is a type of hydraulic axial turbomachinery which has a multistage blade consisting of stators and rotors. In this paper, a turbodrill blade that can be applied in crystallized section under high temperature and pressure conditions is developed. On the basis of Euler equations, the law of energy transfer is analyzed and the output characteristics of turbodrill blade are proposed. Moreover, considering the properties of the layer and the bole-hole conditions, the radical size, the geometrical dimension, and the blade profile are optimized. A computational model of a single-stage blade is built on the ANSYS CFD into which the three-dimensional model of turbodrill is input. In light of the distribution law of the pressure and flow field, the functions of the turbodrill blade are improved and optimized. The turbodrill blade optimization model was verified based on laboratory experiments. The results show that the design meets the deep hard rock mineral exploration application and provides good references for further study.

INTRODUCTION

Turbodrill has been used in oil and gas industry over one century, yet it remains relatively obscure [1, 2]. As geothermal resources, oil and solid mineral [3] are mostly reserved at shallower depths and are now nearly depleted. The decline rate of producing reservoirs accelerates. The search for economically viable reservoirs in the world now focuses on drilling to greater depths. So the drilling technology under high temperature and high pressure (HTHP) conditions has become a hot topic in recent years. Deeper drilling in the crystallized rock formation has faced many challenges such as greater hardness [4], poor formation drillability [5], elevated temperatures [6], higher pressures, and higher costs [7]. All parts of turbodrill are heat-resistant because they are made of metal. Moreover, turbodrill is successful in hard and abrasive formations [8, 9], such as the crystallized rock, because of the compatibility with drill bit types used for drilling and coring these formations and the long life of turbodrills.

Turbodrill is a type of hydraulic axial turbomachinery which has a multistage blade consisting of stators and rotors. It converts the hydraulic power provided by the drilling fluid to mechanical power through turbine motor while diverting the fluid flow through the stator vanes to the rotor vanes [10]. The turbodrill blade is the heart of turbodrill, and its design and casting technologies are very important to the continual success of turbodrill [7]. Compared to the same specifications of PDM (positive displacement motor), the turbodrill rotation speed is usually significantly higher, which leads to the mismatch to the drill bit; the report [11] seeks to build on the successful high temperature turbodrills with additional speed reducer and adjustable bent housing, which are used to drill wells at Los Alamos National Laboratory's hot and dry rock project. Another research [12] introduces a more efficient 2-7/8 in-diameter turbodrill and a novel 4-1/8 in-diameter drill bit for drilling with coiled tubing. Some of the coiled turbine blades [13, 14] are designed on the basis of the requirements of the small diameter drilling. Jianhong et al. [15] calculated the pressure drawdown at the inlet/outlet and the pressure distributions in single-stage turbine for noncoring drilling. In this paper [10], we present computational fluid dynamics (CFD) simulations of a single-stage coiled tube turbodrill performance with different rotation speeds and mass flow rates [16] and then fluid-structural interaction (FSI) analyses for this small size turbodrill in which the finite element analyses of the stresses are performed based on the pressure distributions calculated from the CFD modeling [17]. The turbodrills are widely used in petroleum drilling, but mineral exploration drilling objectives and environment are quite different from petroleum drilling. The core drilling especially is very different from the noncoring drilling with the drilling parameters and drilling process which can lead to particular turbodrill blade.

The turbine drilling tool can be configured to match the application variables as required to optimize performance. Different types of blades can be used to produce different performances [7]. Unfortunately, all of these properties cannot be achieved simultaneously and the optimization always involves some degree of trade-off [18]. In this paper, the basic design methodology of coring turbodrills used in crystallized section is briefly covered. Also the numerical simulation approach for the turbodrill performance analysis is described. Then the simulation results are presented and discussed. At the end, the optimal turbodrill blade is manufactured and tested in laboratory.

HYDRODYNAMIC MODEL OF TURBODRILL BLADE

The Hypotheses of Model

Turbodrill blade converts the hydraulic power provided by the drilling fluid to mechanical power through turbine motor while diverting the fluid flow through the stator vanes to the rotor vanes. Figure 1 shows a typical turbodrill blade assembly and the drilling fluid flow path through turbine stage. The stator vanes are fixed by the axial preload force. Led by the stator vanes (1), the mud flows into the passages among turbodrill blades, leading the rotor (2) to the shaft of the motor which is connected to the bit. (3) The rotors are driven by fluid, providing the required torture and force.

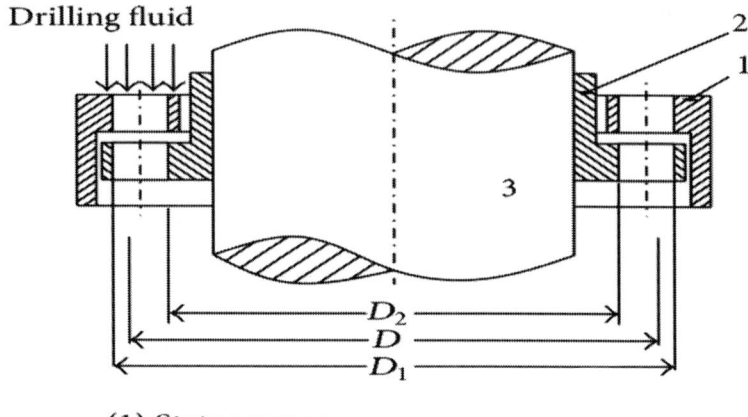

(1) Stator vanes
(2) Rotor vanes
(3) Shaft of the motor

Figure 1: Sketch of the turbodrill blade assembly.

The mud flows into the passages among turbodrill blades which can be regarded as different pieces of coaxial cylindrical movement. Different distances of each layer of liquid particles to the central axis result in different speed. There must be an average diameter D which has the average speed of turbine blade performance which is similar to considering all cylindrical layer flow movement performances of turbine blades. This method is called the unit theory method. So the average diameter is calculated as follows:

$$D = \frac{2\left(D_1^3 - D_2^3\right)}{3\left(D_1^2 - D_2^2\right)}.$$

(1)

On the basis of Euler equations, the hydrodynamic model is built on four hypotheses: (1) the fluid is ideal; (2) turbodrill blade is infinitely thin; (3) turbodrill blade is countless; and (4) the flow and pressure losses can be neglected. So the fluid flow can actually move as expected without any friction.

The Model of Energy Transformation

Figure 2 shows the method of building velocity triangles when analyzing the turbodrill blade unit profile. This method is useful for visualizing changes in the magnitude and direction of the fluid flow due to its interaction with the blade system [3]. The two points 0, points 1, and points 2 along a streamline are the stator vanes input, stator vanes output, and rotor vanes output, respectively. The Bernoulli equation per unit weight between the three points can be obtained as

$$\frac{p_0}{\rho g} + \frac{c_0^2}{2g} + z_0 = \frac{p_1}{\rho g} + \frac{c_1^2}{2g} + z_1 + h_s,$$

$$\frac{p_1}{\rho g} + \frac{c_1^2}{2g} + z_1 = \frac{p_2}{\rho g} + \frac{c_2^2}{2g} + z_2 + h_r.$$

(2)

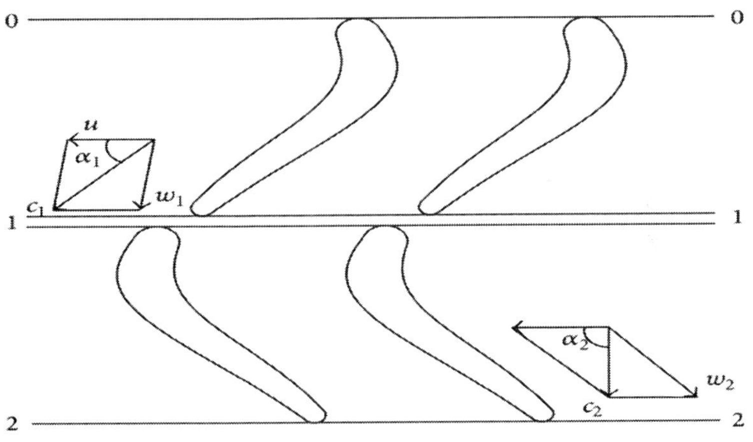

Figure 2: Turbine stage velocity diagrams.

When the fluid flows through the stator vanes, there is no mechanical energy. The pressure energy is transformed into kinetic energy with partial hydraulic losses. However, most energy will be transformed into mechanical energy with partial hydraulic loss when the fluid goes into the rotors vanes.

The different height of drilling fluid can be ignored due to small values between the two sections. Because the flow through each level of the turbine blades is approximately equal and all levels of the length and the structures of the turbine blade are the same, there is no flow law. The flow rate between the turbines at the same level is also similar, meaning that $c_2 = c_0$. So the whole output mechanical energy of single-stage turbine blade is as follows:

$$H_i = \frac{p_0 - p_2}{\rho g} - \left(h_{hs} + h_{hs}\right).$$

(3)

In terms of the single-stage turbine blade, the torque transformed to the rotors is equal to the torque fluid received in different directions. Based on the law of moment of momentum, the output torque of single-stage rotors is calculated as

$$M_i = \frac{1}{2}\rho Q_i D \left(c_1 \cos \alpha_1 - c_2 \cos \alpha_2\right) = \rho Q_i R \left(c_{1u} - c_{2u}\right).$$

(4)

The transfer power consumption of single-stage blade is

$$N_i = M_i \omega = \rho Q_i R \omega \left(c_{1u} - c_{2u}\right) = \rho Q_i u \left(c_{1u} - c_{2u}\right).$$

(5)

According to the energy conservation, the drilling fluid pressure energy can be transformed into mechanical energy. So the energy conversation formula can be expressed as

$$N_i = \rho g Q_i H_i = M_i \omega.$$

(6)

The transformed mechanical energy of the pressure head is

$$H_i = \frac{u}{g}\left(c_{1u} - c_{2u}\right).$$

(7)

The turbodrill blade is a series of the connections in the shell of the turbodrill. Suppose that Z is the numbers of the multistage turbodrill blades. The output torque of single-stage rotors, the energy, and the pressure head are calculated as follows:

$$M_i = Z\rho Q_i R \left(c_{1u} - c_{2u} \right),$$

$$N_i = Z\rho Q_i u \left(c_{1u} - c_{2u} \right),$$

$$H_i = Z\frac{u}{g} \left(c_{1u} - c_{2u} \right).$$

(8)

The Design Model of Turbodrill Blade

The output mechanical energy of turbodrill is related to the numbers of blades stages, radius size, and blade profile. However, the feasible method is increasing the different value of the circumferential speed because of uncomfortable use of the turbine when enlarging the length and radius size of the turbodrill. So it is important to optimize the profile, structure angle, and solidity of the turbodrill blade.

The field conditions must be taken into account when optimizing the turbodrill blade. The value of the circumferential speed of the turbodrill blade changes with the variation of the weight on bit (WOB). Figure 3 displays that a hydraulic loss will occur on the back of the blade exerted when the WOB is big, while the flow separation loss will occur on the basin of the blade when the WOB is small. In order to avoid the loss caused by the impact of fluid, the inlet flow angle $_1$ should be equal to the inlet structure angle $_{1k}$ in stators blade. $_2 = _{2k}$ is also satisfied in rotors blade, in which working condition is called the no-impact status.

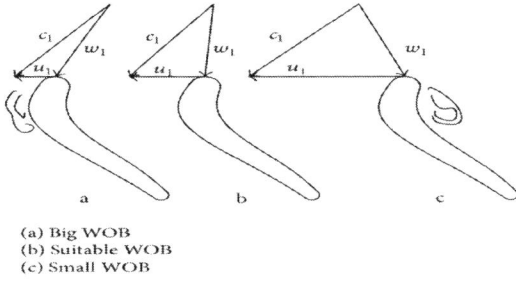

(a) Big WOB
(b) Suitable WOB
(c) Small WOB

Figure 3: Hydraulic loss occurring on different WOB.

Three dimensionless parameters, such as the axial velocity factor $\overline{c_z}$, impact coefficient (m_a), and circulation flow coefficient $\overline{c_u}$, are selected to design and optimize the turbodrill blade profile with fixed size of the stage number and radius size:

$$\overline{c_z} = \frac{c_z}{u_0} = \frac{60Q_i}{\pi^2 D^2 b\varphi n_0} = \frac{1}{\cot \alpha_{1k} + \cot \beta_{1k}}$$

$$= \frac{1}{\cot \alpha_{2k} + \cot \beta_{2k}},$$

$$c_z = \frac{Q_i}{\pi Db\varphi},$$

$$u_0 = \frac{\pi Dn_0}{60}.$$

$$(9)$$

The axial velocity factor is the ratio between the axial flow velocity c_z and the circumferential speed of the blade (u_0), which is related to the blade structure angle. The axial velocity factor is established as (9).

The impact coefficient, which is the ratio of the kinetic energy transformed from the pressure head, is affected only by the blade structure angle and is presented as

$$m_a = \frac{1}{H_i}\left(\frac{c_1^2 - c_2^2}{2g}\right) = \frac{\cot \alpha_{1k} + \cot \alpha_{2k}}{2\left(\cot \alpha_{1k} + \cot \beta_{1k}\right)}.$$

$$(10)$$

The circulation flow coefficient describes the ratio between the dynamic factor and the kinematics indexes; it is affected only by the blade structure angle and is presented as

$$\overline{c_u} = \frac{c_{1u} - c_{2u}}{u_0} = \frac{\cot \alpha_{1k} - \cot \alpha_{2k}}{\cot \alpha_{1k} + \cot \beta_{1k}}.$$

$$(11)$$

Based on the axial velocity factor, the impact coefficient, the circulation flow coefficient, and the structural angle can be deduced as

$$\cot \alpha_{1k} = \frac{c_{1u}}{c_z} = \frac{1}{\overline{c_z}}\left(m_a + \frac{\overline{c_u}}{2}\right),$$

$$\cot \alpha_{2k} = \frac{c_{2u}}{c_z} = \frac{1}{\overline{c_z}}\left(m_a - \frac{\overline{c_u}}{2}\right),$$

$$\cot \beta_{1k} = \frac{\omega_{1u}}{c_z} = \frac{1}{\overline{c_z}}\left[1 - \left(m_a + \frac{\overline{c_u}}{2}\right)\right],$$

$$\cot \beta_{2k} = \frac{\omega_{2u}}{c_z} = \frac{1}{\overline{c_z}}\left[1 - \left(m_a - \frac{\overline{c_u}}{2}\right)\right].$$

(12)

The minimum hydraulic loss occurs when there is no impact on the turbodrill blade. By changing the first derivative differential Euler equation, the output torque, the bit rotation speed, the turbodrill blade pressure drop, and the running-in speed can be calculated as

$$M_0 = \frac{\eta_v^2 \eta_m}{2\pi b \varphi} \rho Q^2 \frac{\overline{c_u}}{c_z},$$

$$n_0 = \frac{60\eta_v}{(\pi D)^2 b \varphi} \frac{Q}{\overline{c_z}},$$

$$H_0 = \left(\frac{\eta_v}{\pi D b \varphi}\right)^2 \rho \frac{Q^2}{\eta_h} \frac{\overline{c_u}}{\overline{c_z}^2},$$

$$n_x = n_0 \cdot \left(1 + \overline{c_u}\right).$$

(13)

The equation shows that the output torque under the premium working condition is proportional to circulation flow coefficient. The bit rotation speed is only opposite to the axial velocity factor. The specialized function of the turbodrill blade can be translated by adjusting the axial velocity factor and circulation flow coefficient.

When the hydraulic efficiency η_h is increasing, the pressure drop will be reducing simultaneously. The hydraulic efficiency is also related to the axial velocity and circulation flow coefficient. However, considering the hydraulic efficiency, it is recommended that the axial velocity factor and circulation flow coefficient should be adjusted within proper domain.

DESIGN OF TURBODRILL BLADE

The Design Goal and Diagram

The turbodrill is designed to satisfy the coring in the crystalline rock formations, whose drillability classification number is from 7th to 8th. The parameters of the turbodrill blade should be designed as in Table1.

Table 1: Target performance parameters of turbodrill ($\Phi 127$)

Items/unit	Value
Outside diameter/mm	127
Working flow/L·s^{-1}	10~15
Rotation speed/r·min^{-1}	200~500
Pressure drop/MPa	≤5
Rated torque/N·m	800~1200
Drilling fluid density/ kg·m^{-3}	1000~2000

The turbine blade design includes the parameters of the working characteristics, the optimum radius, the axial dimension, and the profile and the parameters of the blade structures. The design diagram for the turbodrill blade is shown in Figure 4. First, the working characteristics, such as the rated output torque, the optimal rotation speed, and the pressure drop, are defined by the coring demand in the crystalline rock formations.

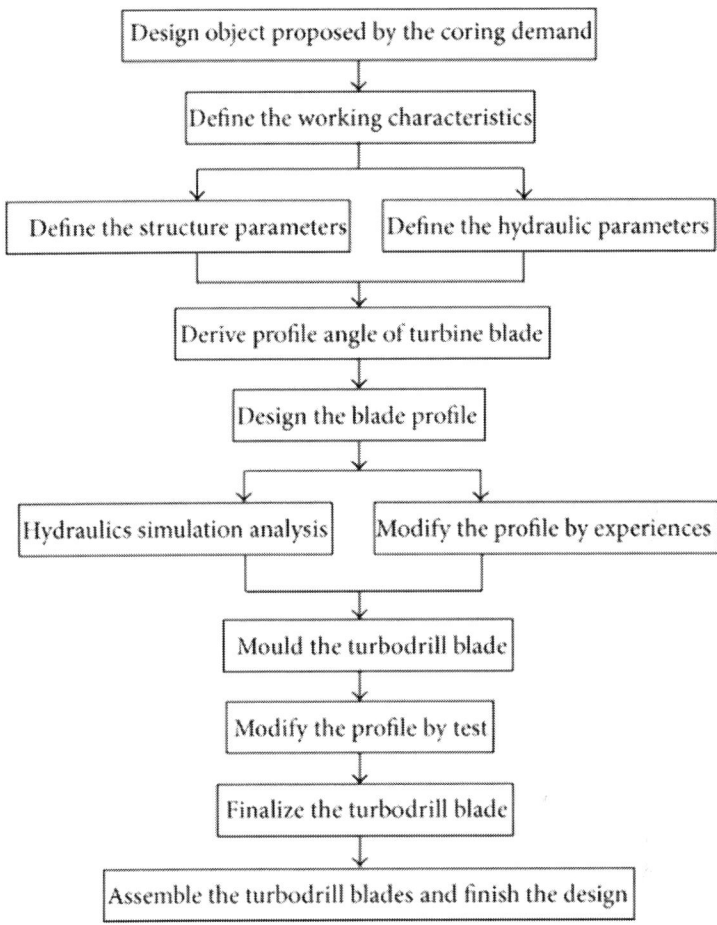

Figure 4: The design diagram for turbodrill blade.

The outer diameter Φ 127 mm turbodrill is generally equipped with Φ 152 mm impregnated diamond bit. 2–4 rpm of linear velocity ensures a higher rock-crushing efficiency. The range of 250–500 rpm is regarded as the optimal velocity after our calculation, demanding that the blade structure should be adjusted to lower the speed. Second, after the structure parameters are preliminary determined, such as the radius and the axial dimension, the parameters of blade structure angles can be calculated under the no-impact status according to three precedent dimensionless parameters. So the turbodrill profiles are drawn out. Third, they are optimized repeatedly by the simulations, the experience, and the laboratory experiments until the output performances of turbodrill are satisfied.

Structure Design of Turbodrill Blade

Except for the profile, the outer cylinder size, the radius size, and the axial dimension, other parts are also included in the design. The calibration and material selection are also accomplished in this section. The outer diameter of the tube and turbine stator should be increased as big as possible according to the borehole and annular clearance. The material of the turbodrill tube is 40CrMo alloy steel which has high mechanical properties and good performances of heat treatment for petroleum drilling.

The solid rotor shaft, which bears the positive torque from the rotors and antitorque from the bit, should be as slim as possible to meet sufficient stress intensity. The relationship between the diameter of the rotor shaft and the allowable twisting stress is shown as in (14). The diameter of the rotor shaft is equal to the inner diameter of the rotor with a clearance fit, so the inner diameter of the turbine rotor is decided by the rotor shaft:

$$d \geq \sqrt[3]{\frac{ST}{[\tau]}}.$$

(14)

In order to avoid the interference between stator vanes and rotor vanes, a clearance should be retained. Considering the precision machining limit, the clearances between the static and dynamic component parts are about 1 to 2 mm. According to this design principle, the parts of the design dimensions are shown in Figure5.

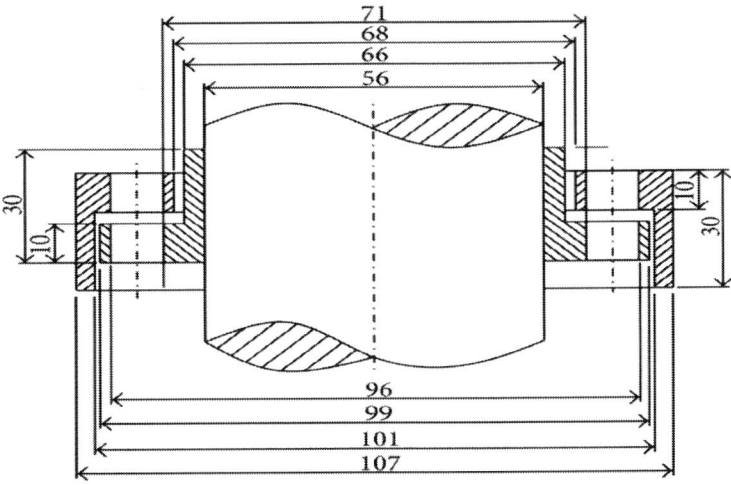

Figure 5: Part of the design dimensions.

When the volumetric efficiency η_v is 0.9 and the reduction coefficient of the blade φ is 0.9, the axial component of the absolute flow velocity is calculated as

$$C_{1z} = \frac{Q \cdot \eta_v}{\pi D b \varphi} = (10 \sim 15) \cdot 0.287 \, \text{m/s} = (2.87 \sim 4.315) \, \text{m/s.} \tag{15}$$

Profile Design of Turbodrill Blade

The blade runner field is changed mainly by influencing the speed and pressure of the flow near the blade surface. In order to keep a higher efficiency, the speed and pressure should be changed smoothly, reducing the fluid friction, flow separation loss, and blade end pressure loss. Since then, 2D orthogonal curvilinear coordinates are established to research the influence of the blade line on the law of the fluid move on the blade surface. The fluid field can be changed by affecting its speed and pressure. That smooth speed and the pressure distribution ensure high efficiency. Figure 6 displays a 2D coordinate of the turbodrill blade boundary. On the coordinates of the point P(x,y), x is the dorsal arc along the surface profile from the origin (O), and y is the vertical distance from the normal line of the surface profile. u and v are the corresponding speeds of the two directions, respectively.

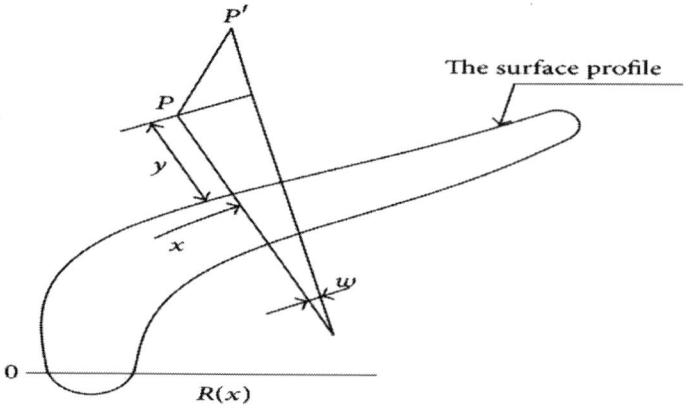

Figure 6: The 2D coordination of the turbodrill blade boundary.

On the basis of the primary motion equation about the steady flow boundary of the incompressible fluid, the massive conservation and momentum conservation can be expressed as

$$\frac{\partial u}{\partial x} + \frac{\partial}{\partial x}\left[\left(1 + \frac{y}{R}\right)v\right] = 0,$$

$$\frac{Ru\partial u}{(R+y)\partial x} + v\frac{\partial u}{\partial y} + \frac{uv}{R+y} = -\frac{R\partial P}{(R+y)\rho\partial x}$$

$$+ v\left[-\frac{R\partial^2 v}{(R+y)\partial x\partial y} + \frac{\partial^2 u}{\partial y^2}\right.$$

$$+ \frac{\partial u}{(R+y)\partial y} + \frac{R\partial v}{(R+y)\partial x} - \frac{u}{(R+y)^2}\Big],$$

$$\frac{Ru\partial v}{(R+y)\partial x} + v\frac{\partial v}{\partial y} - \frac{u^2}{R+y} = -\frac{\partial P}{\rho\partial y}$$

$$+ v\left[-\frac{R\partial^2 u}{(R+y)\partial x\partial y} - \frac{R\partial u}{(R+y)^2\partial x} + \frac{R^2\partial^2 v}{(R+y)^2\partial x^2}\right.$$

$$+ \frac{R}{(R+y)^2}\frac{dR}{dx}\left[\frac{u}{R+y} + \frac{y\partial u}{(R+y)\partial x}\right]\right].$$

(16)

The equations indicate that boundary curvature radius (R) influences the fluid parameter most. A continuous derivative of curvature can ensure smooth distribution. Suppose that the profile of the turbodrill blade is y=f(x) and its curvature and third derivative are, respectively,

$$C = \frac{1}{R} = \frac{y''}{\left[1 + (y')^2\right]^{3/2}} = \frac{f''}{\left[1 + (f')^2\right]^{3/2}},$$

$$C = \frac{f''' \left[1 + (f')^2\right] - 3f'(f'')^2}{\left[1 + (f')^2\right]^{5/2}}.$$

(17)

Although combined profile can meet continuous shrinkage, it has no continuous curvature derivative. This can cause sudden change of the fluid speed and pressure and damage hydraulic property of the turbodrill. As a result, we use quintic polynomial combined with computer-aided design to develop profile of the turbodrill. Suppose that the pressure side and negative pressure side of the blade are $y_p = f(x)$ and $y_s = g(x)$. Consider

$$y_p = a_0 + a_1 x + a_2 x^2 + a_3 x^3 + a_4 x^4 + a_5 x^5,$$

$$y_s = b_0 + b_1 x + b_2 x^2 + b_3 x^3 + b_4 x^4 + b_5 x^5.$$

(18)

Based on the angles of the inlet and outlet, the four special points can be calculated. Put the four special points, first derivative, and second derivative into their equations, and the pressure side and negative pressure side of the blade can be described as

$$\begin{bmatrix} 1 & x_{p1} & x_{p1}^2 & x_{p1}^3 & x_{p1}^4 & x_{p1}^5 \\ 1 & x_{pn} & x_{pn}^2 & x_{pn}^3 & x_{pn}^4 & x_{pn}^5 \\ 0 & 1 & 2x_{p1} & 3x_{p1}^2 & 4x_{p1}^3 & 5x_{p1}^3 \\ 0 & 1 & 2x_{pn} & 3x_{pn}^2 & 4x_{pn}^3 & 5x_{pn}^3 \\ 0 & 0 & 2 & 6x_{p1} & 12x_{p1}^2 & 20x_{p1}^3 \\ 0 & 0 & 2 & 6x_{pn} & 12x_{pn}^2 & 20x_{pn}^3 \end{bmatrix} \begin{bmatrix} a_0 \\ a_1 \\ a_2 \\ a_3 \\ a_4 \\ a_5 \end{bmatrix}$$

$$= \begin{bmatrix} y_{p1} \\ y_{pn} \\ y'_{p1} \\ y'_{pn} \\ y''_{p1} \\ y''_{pn} \end{bmatrix},$$

(19)

$$\begin{bmatrix} 1 & x_{s1} & x_{s1}^2 & x_{s1}^3 & x_{s1}^4 & x_{s1}^5 \\ 1 & x_{sn} & x_{sn}^2 & x_{sn}^3 & x_{sn}^4 & x_{sn}^5 \\ 0 & 1 & 2x_{s1} & 3x_{s1}^2 & 4x_{s1}^3 & 5x_{s1}^4 \\ 0 & 1 & 2x_{sn} & 3x_{sn}^2 & 4x_{sn}^2 & 5x_{sn}^2 \\ 0 & 0 & 2 & 6x_{s1} & 12x_{s1}^2 & 20x_{s1}^3 \\ 0 & 0 & 2 & 6x_{sn} & 12x_{sn}^2 & 20x_{sn}^3 \end{bmatrix} \begin{bmatrix} b_0 \\ b_1 \\ b_2 \\ b_3 \\ b_4 \\ b_5 \end{bmatrix}$$

$$= \begin{bmatrix} y_{s1} \\ y_{sn} \\ y'_{s1} \\ y'_{sn} \\ y''_{s1} \\ y''_{sn} \end{bmatrix} \cdot$$

$$\tag{20}$$

The parameters of a_0, a_1, a_2, a_3, a_4, a_5, b_0, b_1, b_2, b_3, b_4, and b_5 can be solved by the two linear equations (19) and (20). Figure 7 displays the profile of the pressure side and negative pressure side of the blade when the output rotation speed n_0=500 rpm and M_0=1000 output torque Nm which are given by the corresponding relationship between torque and speed.

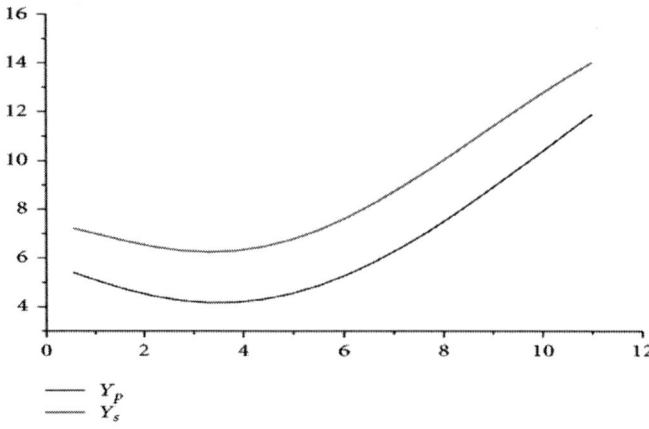

Figure 7: The preliminary profile of the blade.

After checking the blade line, it shows that there are no reflection points on pressure surface and suction surface. This constitutes a continuous flow path contraction. It proves that the line has a relatively flow loss, meeting all the design requirements. Finally, according to the axial length of size constraints, a leading edge circle and a trailing edge circle are established on its tangent line at both ends of the blade so as to make the blade closed.

CFD SIMULATION OF TURBODRILL BLADE

CFD Model

Computational fluid dynamics (CFD) has emerged as an effective optimization tool for the experiments because of its diverse applications in industry [19, 20]. Drawing with the SolidWorks a single cycle turbine blade airfoil-dimensional model (Figure 8), we can establish the flow field width on the basis of blade profile in two-dimensional model, extend three times of upper and lower widths to obtain import and export flow borders, and form a closed fluid channel; it is easy to get the stable solution of flow field and establish a single cycle cross-flow model, and we can establish complete three-dimensional models in SolidWorks formed inANSYS CFD simulation model.

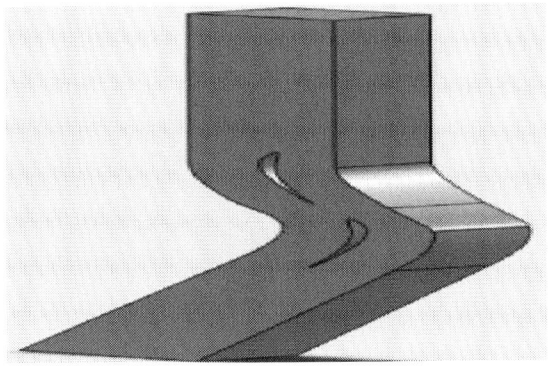

Figure 8: The single-period model in SolidWorks.

Meshing and Boundary Conditions

The blade is characterized by irregular shape, demanding an automatic mesh generation. In this simulation, a minimum unit meshing by free division ensures accurate results. There are 142 units meshing the finite element model in Figure 9. Although significantly greater than the extension of the grid around the blade grid, the meshes near the blade are relatively uniform. This proves that the mesh quality is high. ANSYS is used to research the differences between the flow fields. The curve of the blade is complex—short runner and relatively fast flow rate. After calculating Reynolds (Reynolds number), it proves to be turbulent. Different models need to set different solution control and execution control. Because this three-dimensional simulation of flow field is continuously differentiable turbulent flow, the calculation of control theory should be turbulence model theory. The default equation is Reynolds averaged equations. For unsteady turbulent flow, ANSYS analysis cites false concept of homeostasis without regard to the circumferential direction of the flow field changes, so at this mode of analysis it should be TRAN steady flow. The input flow and output flow are stable when the blades are applied, resulting in flow between the stator and the rotor blades relatively stable.

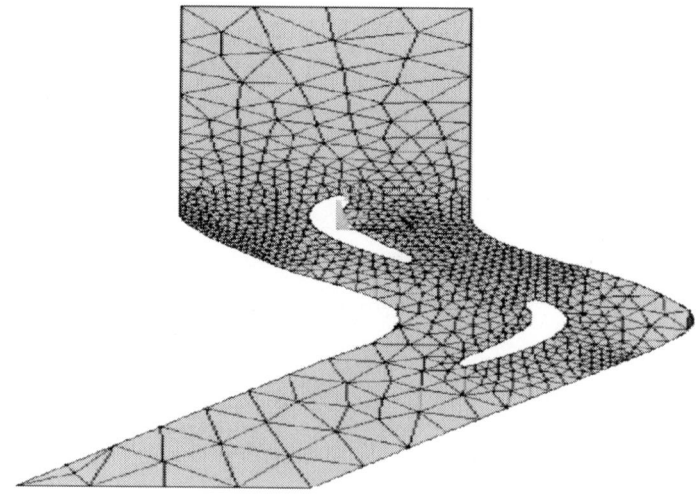

Figure 9: The ANSYS model of the turbodrill blade.

For the drilling fluid of the turbine blade, with certain density and viscosity, we can select fluid and then analyze and define these two worthy constants when we set fluid conditions. Other parameters generally use the defaults of ANSYS program system. After completing the relevant definitions and settings, we use Run Flotran to solve the corresponding analysis of the results. In the ANSYS postprocessor tune the corresponding calculation results are obtained using the PLOT function blade flow velocity and pressure fields for analysis.

Simulation Results

In order to select an optimum blade shape and verify its effectiveness, the CFD analysis of the two different turbine blades on the basis of the preliminary profiles was completed for low speed high torque flow field simulation. Figures 10 and 11 display the simulation results, including the pressure field of the blade, Y-component of fluid velocity, X-component of fluid velocity, Z-component of fluid velocity, fluid velocity, and turbulent energy dissipation.

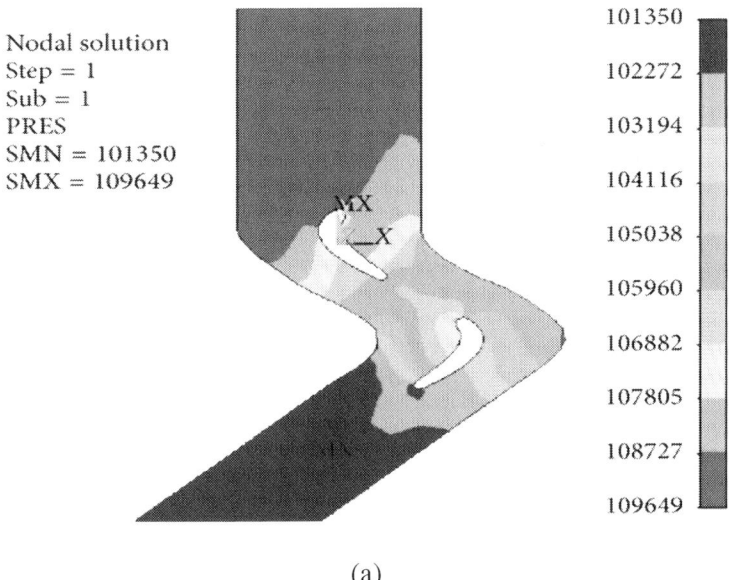

(a)

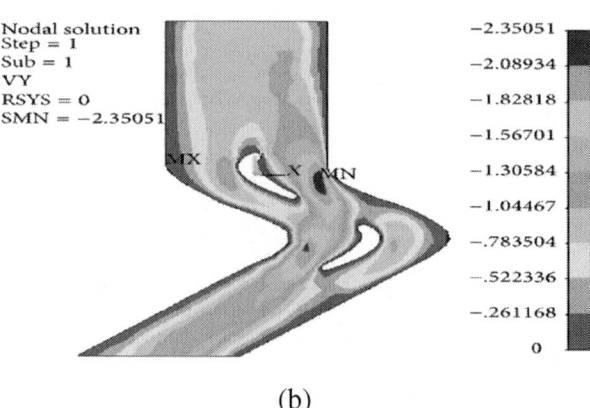

(b)

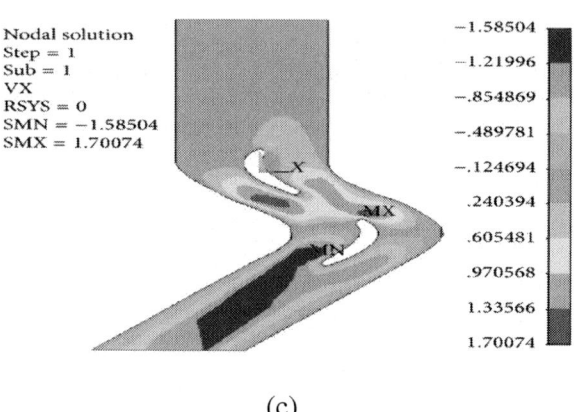

(c)

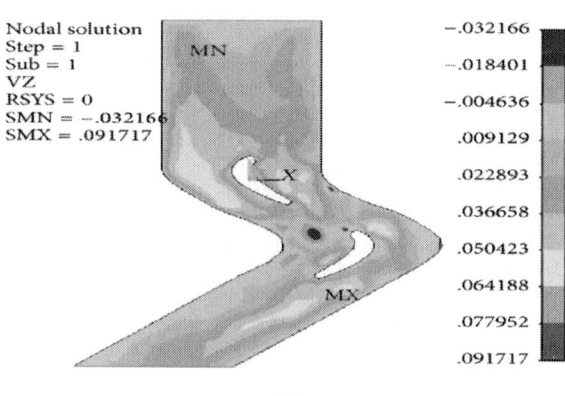

(d)

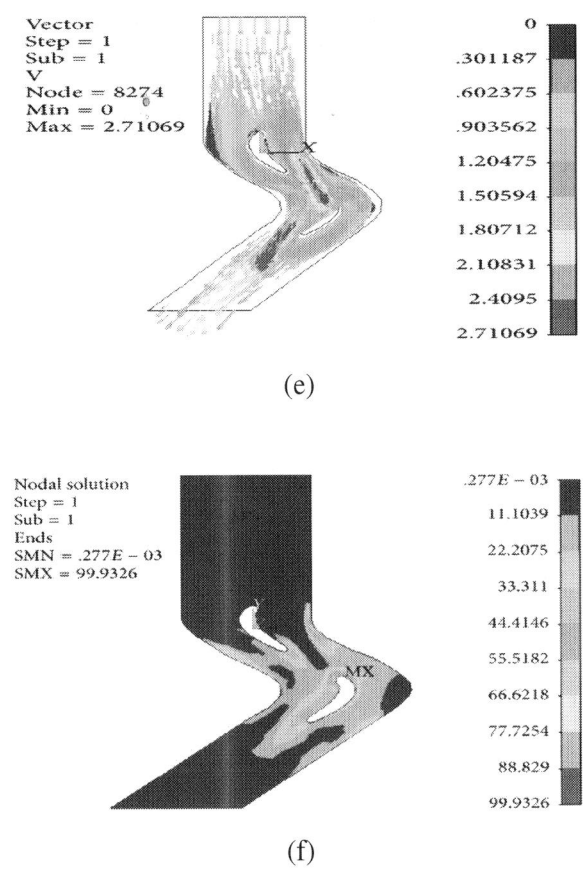

(e)

(f)

Figure 10: The flow field of I-type blade.

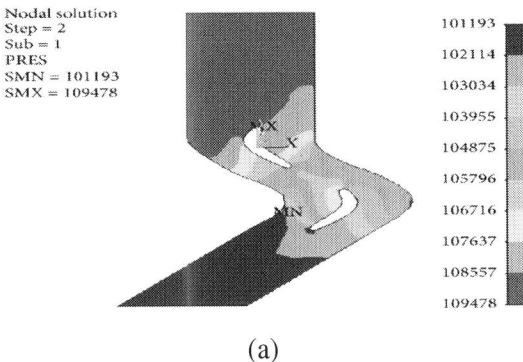

(a)

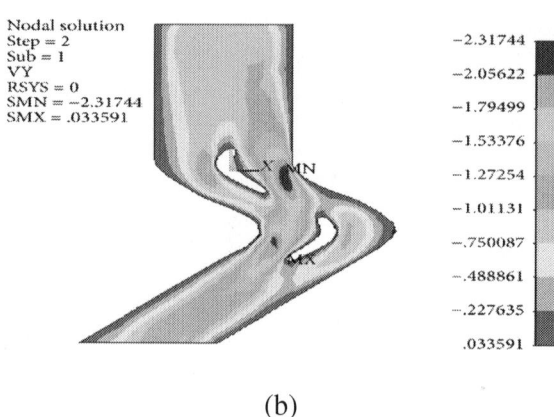

(b)

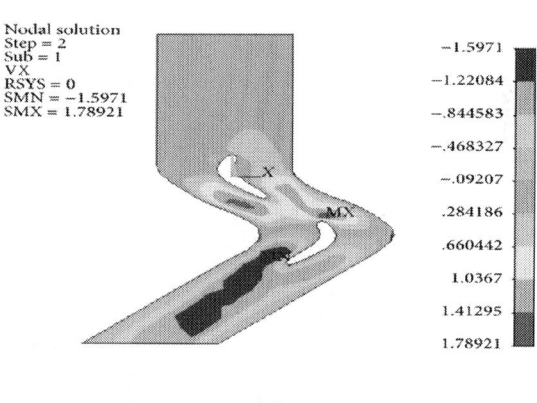

(c)

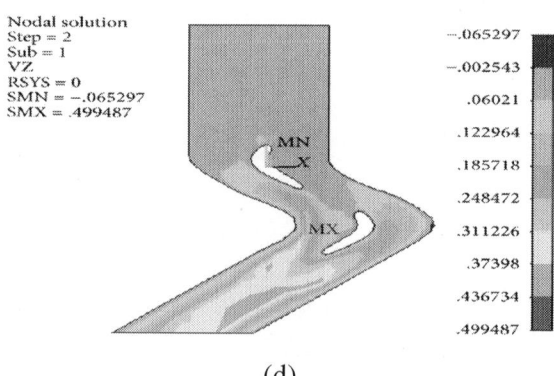

(d)

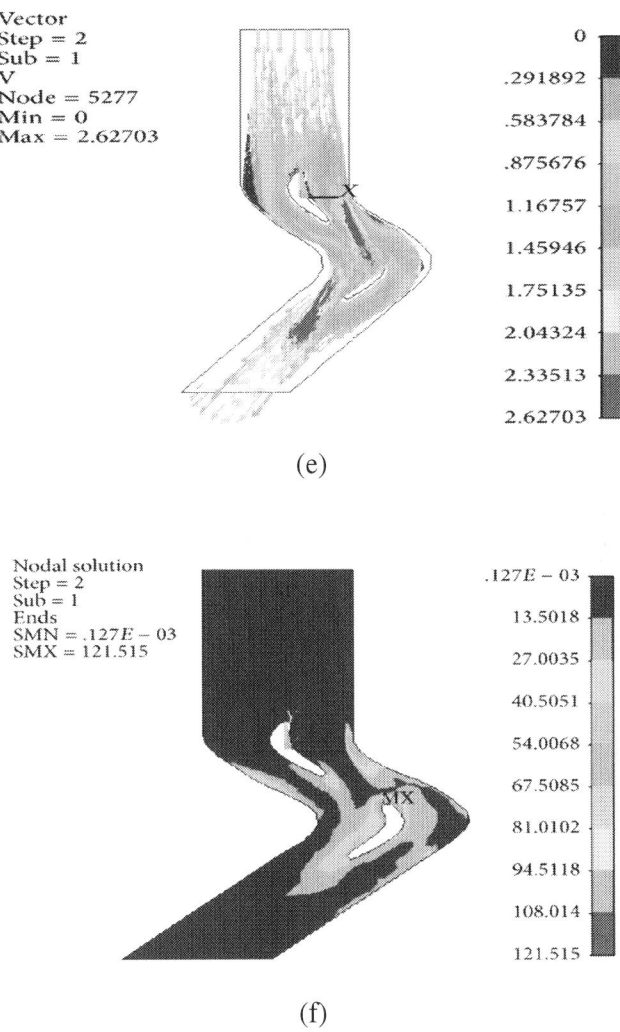

(e)

(f)

Figure 11: The flow field of II-type blade.

The flow field simulation figure shows the following. (1) The Y-component and X-component of fluid velocity are the drilling fluid dominant orientation. These two kinds of fluid velocity become larger smoothly and gradually, which proves the reasonable design of the surface profile. (2) The Z-component of fluid velocity of both types of blade is very small, which shows the little energy dissipation. The Z-component of fluid velocity of the I-type blade is smaller than the II-type blade, which indicates that the former has less energy dissipation. (3) The drilling fluid flows into the turbine blades of the stator. The fluid impacts pressure surface and the suction surface after flowing into the blade. Then it will accelerate along the flow channel. The rotor work flow and pressure energy can be translated into kinetic energy of the rotor mechanical kinetic energy. (4) The pressure field surface pressure distribution from the inlet to the outlet ends in smooth descending order. The turning point of the leading edge of the blade is greater than the blade suction, from the description of the pressure gradient of the output torque of the size. The greater pressure gradient of the I-type leaf blade shows that it has more torque gradient than the II-type blade. (5) The small turbulent energy dissipation conforms to the Z-component of fluid velocity. The turbulent energy dissipation of the I-type blade is smaller than the II-type blade, which indicates that the former is better. In summary, the overall smooth flow and the low eddy current loss prove that both types of turbine blades correction are feasible. The results meet all the requirements. After careful comparative analysis, the I-type blade turns out to be more reasonable and suitable to manufacture the physical prototype for test.

TEST OF THE TURBODRILL STAGE

Turbodrill Blade Casting

Turbodrill blade is processed using fine casting molding. First, based on the blade structure and its curvature, a fine model (Figure 12(a)) is processed by CNC. Second, a vice blade blank should be modeled by pouring wax, shown in Figure 12(b). After the mating surface is processed, a blade can be made (Figure 12(c)).

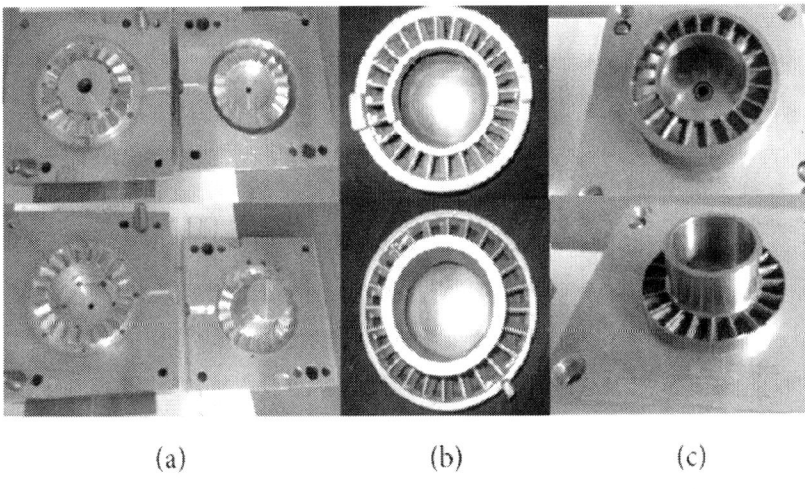

(a)   (b)   (c)

Figure 12: The casting process of the turbodrill blade.

Laboratory Test of the Turbodrill Blade

The turbodrill prototype is composed of two turbine sections, which has one hundred turbine blades. In order to verify the effect of the turbodrill coring blade design, the single turbine section was tested on the drill test bench in Tianjin Dagang mechanical manufacturing companies on July 14, 2013. As shown in Figure 13, the test rig is mainly composed of the host, load system, circulatory system, and measurement system [21]. Experimental principle is as follows: the volume flowing into the turbine drill traffic is kept to a given value under the premise of the drill through the turbine power output of the torque applied to the different loads, so that drilling in different stable braking torque works. It can test output torque, output speed, and pressure loss under different conditions. Also, it can be used to research the relationship among the output torque, efficiency of the turbine drill, and the rotational speed.

Figure 13: Laboratory test and its equipment.

Figure 14 shows the output torque and power variation with the rotational speed. By using the water (1 g/cm³) as the drilling fluid, the maximal output torque and maximal rotational speed of the single turbine section are 394 N·m and 564 rpm, respectively. Considering the actual density of the drilling fluid is the 1.25–1.3 times of water and that the output torque is proportional to the density of the drilling fluid, the maximal output torque of the turbodrill prototype with two turbine sections can reach up to 985–1025 N·m, which is equal to the design goal. With the increase of the drilling fluid, the rotational speed will decrease slightly. The test results are equal to the design goals, which can verify the accuracy in surface profile indirectly.

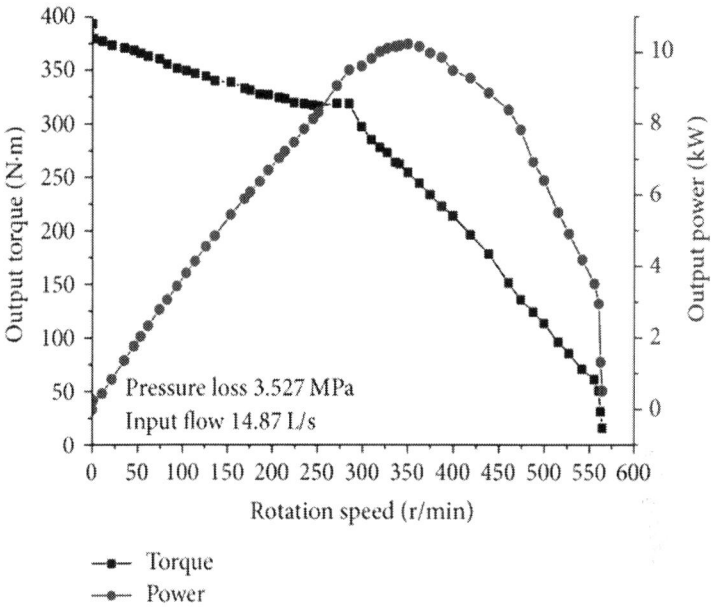

Figure 14: Laboratory test results.

CONCLUSIONS

- Considering high pressure and temperature, hydraulic units with high torque and low speed are applied. The turbodrill is developed with optimized parameters. The surface profiles of the turbodrill are designed on the basis of dimensionless coefficients.
- The effect of the blade profile's change on the first curvature is figured out: the first curvature of the blade front virtually affects the blade front end. Lower curvature leads to higher front and smoother flat. First curvature of the blade back virtually affects the blade back end. Lower curvature leads to higher turbodrill blade profile.
- The single-period model is developed through ANSYS CFD. The effects of different fluid discharge and viscosity on hydraulic property are performed. The optimized design fitting curve is confirmed.

- The basic design methodology and method of coring turbodrills used in crystallized section are efficacious. The results show that the design meets the deep hard rocks mineral exploration application and provides good references for further study.

ACKNOWLEDGMENTS

The authors gratefully acknowledge the support by the International Scientific and Technological Cooperation Projects (Grants nos. 2010DFR70920 and 2011DFR71170), the Beijing Organization Department Outstanding Talented Person Project (no. 2013D009015000002), and the Beijing Higher Education Young Elite Teacher Project (Grant no. YETP0645). Meanwhile, great thanks also go to former researchers for their excellent works, which gave great help for the authors' academic study.

REFERENCES

1. U.S. Department of Energy, "Former USSR R&D on advanced downhole drilling motors," Russian Drilling Technology Studies (CD-ROM) E 1.111:R92, 1997.

2. F. V. Delucia, "Benefits, limitations, and applicability of steerable system drilling," in Proceedings of the SPE/IADC Drilling Conference, SPE-18656-MS, New Orleans, La, USA, February 1989. ·

3. A. Mokaramian, V. Rasouli, and G. Cavanough, "Adapting oil and gas downhole motors for deep mineral exploration drilling," in Proceedings of the 6th International Seminar on Deep and High Stress Mining, Perth, Australia, 2012.

4. L. Guibin, L. Yongjing, and L. Yaoquan, "Bit selection and application for granite rock of Well Pugu-1,"Oil Drilling & Production Technology, vol. 33, no. 6, pp. 106–109, 2011.

5. X. Baoping, Z. Jinchang, and Z. Yangsheng, "Key technologies of hot dry rock drilling during construction," Chinese Journal of Rock Mechanics and Engineering, vol. 30, pp. 2234–2243, 2011.

6. Y. Wang, L. Bao-lin, Z. Hai-yan, et al., "Thermophysical and mechanical properties of granite and its effects on borehole stability in high temperature and three-dimensional stress," The Scientific World Journal, vol. 2014, Article ID 650683, 11 pages, 2014. ·

7. D. Calnan, R. Seale, and T. Beaton, "Identifying applications for turbodrilling and evaluating historical performances in North America," Journal of Canadian Petroleum Technology, vol. 46, no. 6, pp. 34–39, 2007.

8. T. Chunfei, W. Jiachang, and Z. Weijian, "Test of reduction turbodrill TDR1-127 at ultra-deep wells in Tahe oilfield," Oil Drilling & Production Technology, vol. 32, pp. 93–96, 2010.

9. X. Zhang and J. Feng, "Research on the effect of fluid media on the mechanical property of high speed turbo-drill," China Pertroleum Machinery, vol. 40, pp. 38–42, 2012.

10. A. Mokaramiani, V. Rasouli, and G. Cavanough, "CFD simulations of turbodrill performance with asymmetric stator and rotor blades congiguration," in Proceedings of the 9th International Conference on CFD in the Minerals and Process Industries, CSIRO, Melbourne, Australia, December 2012.

11. W. C. Maurer, Advanced Geothermal Turbodrill, Technology International, US Department of Energy, National Energy Technology Laboratory (NETL), 2000.

12. R. Radtke, D. Glowka, M. M. Rai, T. Beaton, and R. Seale, High-Power Turbodrill and Drill Bit for Drilling with Coiled Tubing, Technology International, US Department of Energy, National Energy Technology Laboratory (NETL), 2011.

13. R. Seale, T. Beaton, and G. Flint, "Optimizing turbodrill designs for coiled tubing applications," inProceedings of the SPE Eastern Regional Meeting, Charleston, WV, USA, 2004.

14. C. Grigor, D. Conroy, and M. Henderson, "Expanding the use of turbodrills in coiled tubing and workover applications," in Proceedings of the SPE/ICoTA Coiled Tubing and Well Intervention Conference and Exhibition, SPE-113721-MS, The Woodlands, Tex, USA, April 2008. ·

15. F. Jianhong, S. Kexiong, and Z. Zhi, "Analysis on three-dimensional numerical simulation of turbine flow characteristics," Energy Procedia, vol. 16, pp. 1259–1263, 2012. ·

16. A. Mokaramian, V. Rasouli, and G. Cavanhough, "A feasibility study on adopting coiled tubing drilling technology for deep hard rock mining exploration," in Proceedings of the 6th International Seminar on Deep and High Stress Mining, Perth, Australia, 2012.

17. A. Mokaramian, V. Rasouli, and G. Cavanough, "Fluid flow investigation through small turbodrill for optimal performance," Mechanical Engineering Research, vol. 3, no. 1, pp. 1–24, 2013.·

18. J. R. Bourgoyne, A . T. Millheim, K. K. Chenevert, M. E. Chenevert, and F. S. Young, Applied Drilling Engineering, vol. 2 of SPE Textbook Series, Society of Petroleum Engineers, Richardson, Tex, USA, 1991.

19. M. Keerthana and P. Harikrishna, "Application of CFD for assessment of galloping stability of rectangular and H-sections," Journal of Scientific and Industrial Research, vol. 72, no. 7, pp. 419–427, 2013.

20. S. N. Singh, V. Seshadri, R. K. Singh, and T. Mishra, "Flow characteristics of an annular gas turbine combustor model for reacting flows using CFD," Journal of Scientific and Industrial Research, vol. 65, no. 11, pp. 921–934, 2006.

21. Q. Zhang, M. Li, Z. Cheng, et al., "Development and application of test bed for large torque screw drill,"China Petroleum Machinery, vol. 35, no. 7, pp. 31–34, 2007.

Chapter 3

Wellbore Stability in Oil and Gas Drilling with Chemical-Mechanical Coupling

Chuanliang Yan[1,2], Jingen Deng[1], and Baohua Yu[1]

[1]State Key Laboratory of Petroleum Resource and Prospecting, China University of Petroleum, Beijing 102249, China

[2]Department of Petroleum Engineering, China University of Petroleum, Beijing 102249, China

ABSTRACT

Wellbore instability in oil and gas drilling is resulted from both mechanical and chemical factors. Hydration is produced in shale formation owing to the influence of the chemical property of drilling fluid. A new experimental method to measure diffusion coefficient of shale hydration is given, and the calculation method of experimental results is introduced. The diffusion coefficient of shale hydration is measured with the down hole temperature and pressure condition, then the penetration migrate law of drilling fluid filtrate around the wellbore is calculated. Furthermore, the changing rules of shale mechanical properties affected by hydration and water absorption are studied through experiments. The relationships between shale mechanical parameters and the water content are established. The wellbore stability model chemical-mechanical coupling is obtained based on the experimental results. Under the action of drilling fluid, hydration makes the shale formation softened and produced the swelling strain after drilling. This will lead to the collapse pressure increases after drilling. The study results provide a reference for studying hydration collapse period of shale.

INTRODUCTION

Maintaining wellbore stability is an important issue in oil and gas industry [1–10]. In the process of drilling, the economic losses caused by wellbore instability reaches more than one billion dollar every year [11], and the lost time is accounting for over 40% of all drilling related nonproductive time [12]. It is also reported that shale account for 75% of all formations drilled by the oil and gas industry, and 90% of wellbore stability problems occur in shale formations [13–18]. When a well is drilled, the formation around the wellbore must sustain the load that was previously taken by the removed formation. As a result, an increase in stress around the wellbore and stress concentration will be produced [19–23]. If the strength of the formation is not strong enough the wellbore will be failure [24–28]. Wellbore stability is not only a pure rock mechanical problem, but also the interaction of drilling fluid and shale is a more important influence factor [29–35]. There are various chemicals in the drilling fluid which physically and chemically interact with shale formations. One hand, these interactions will result in the production of swelling stress [36–43]. On the other hand, it alleviates the mechanical strength of the wellbore wall rock [44–46]. Furthermore, it results in wellbore instability.

When studying the wellbore stability in shale, chemical factor must be combined with mechanical factor. Before the 1990s, the combinations are mainly on experimental study. Chenevert studied mechanical properties of shale after hydration since 1970s [44]. The results showed that the hydration would decrease the shale strength. After 1990s, the combinations came into a quantitative research stage. Yew et al. (1990)

[29] and Huang et al. (1995) [47] combined shale hydrated effect quantitatively into the mechanical model based on thermoelasticity theory. Their method attributed the rock mechanical properties change with total water content. Take shale as a semipermeable membrane, Hale et al. (1993) [48, 49], Deng et al. (2003) [50], and Zhang et al. (2009) [51] introduced equivalent pore pressure to study interaction of shale and water base drilling fluid. Ghassemi et al. (2009) [52] proposed a linear chemo-thermo-poroelasticity coupling model, which considers the influence of chemical potential and temperature. Wang et al. (2012) [53, 54] built a fluid-solid-chemistry coupling model, in which they considered electrochemical potential, fluid flow caused by ion diffusion.

The chemical effect of drilling fluid on shale can be ultimately attributed to the variation of rock mechanical properties and stress around the wellbore. Water migration in shale is the basement of all wellbore stability models with chemical-mechanical coupling. A new experimental equipment to measure in situ water diffusion coefficient of shale is developed in this paper. And a sample model to evaluate time-dependent collapse pressure with chemical-mechanical coupling is presented.

EXPERIMENTAL RESEARCH ON THE HYDRATION OF SHALE

The free water and ion will penetrate into shale under the driving force of chemical potential and pressure difference between the pore fluid and drilling fluid [55–58]. Water content of shale changes by various mechanisms such as osmosis flow, viscous flow and capillary flow. Osmosis flow, driving force is due to chemicals and ions with different composition in drilling fluid and pore fluid. In order to evaluate the hydration of drilling fluid, the coefficient of water absorption and diffusion and the swelling ratio must be determined first [36].

Experimental Research on the Water Absorption of Shale

Experimental Equipment

Cherevent let one end face of shale sample contact with drilling fluid and the other end face wrapped up by plastic film, then he measured the water content increment in different location. But his experiment can only be conducted in room temperature and with zero confining pressure. But during drilling process in deep formation, it is in the condition of high temperature and high pressure. Shale hydration is influenced by temperature and pressure seriously, so his experimental result was inconsistent with actual drilling. In order to test the coefficient of water diffusion of shale, we developed an in situ test equipment of water diffusion coefficient which can fit the down whole temperature and pressure condition while drilling (Figure1).

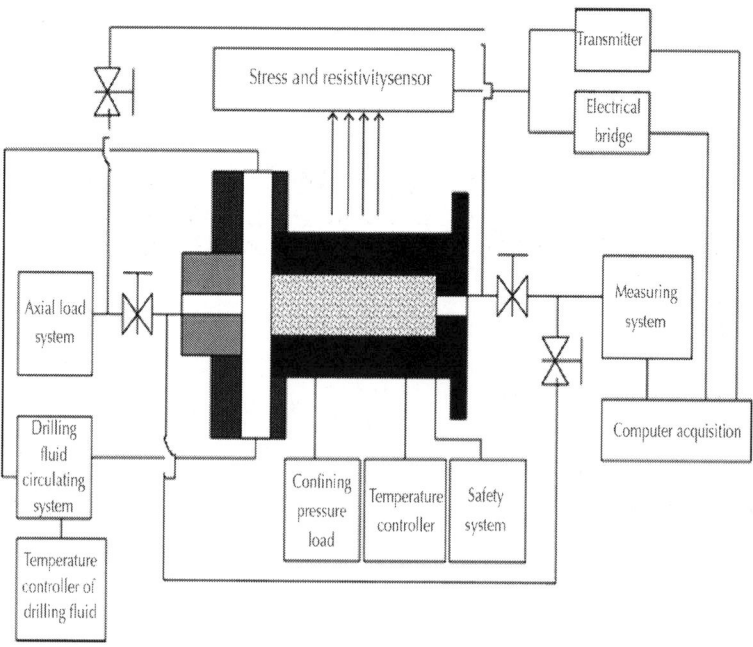

Figure 1: The experimental equipment sketch.

Technical parameters of this designed equipment are as the following.

- Temperature: room temperature to 150°C, which can imitate the temperature condition of the formation with 5000 meters depth.
- Pressure: confining pressure 0 MPa to 70 MPa, axial pressure 0 MPa to 200 MPa.
- Imitate the maximum differential pressure of drilling fluid with 10 MPa.
- Sample size: ⌀ 25 mm × 50 mm.

The experimental process are as follows.

- Determine the original water content of the rock samples first, wrap the samples with separation sleeve, and put into the core holder. Put the drilling fluid into the tank and check the test system to make sure it is in good condition.
- Turn on the temperature controller, warm the core samples to the same temperature with down whole condition. Then load the confining pressure and axial pressure to proper value and start timing.
- During the test, data acquisition control system is used to keep the test values constant.
- Cooling uninstall when the test time reaches the predetermined value (50 hours in this research), remove the rock samples quickly, and measure the water content at different distance from the end face.

Coefficient of Water Diffusion

According to conservation of mass, water diffusion equations can be established. Supposing q is mass flow rate of the water diffusion, C_s (r,t) is the weight percentage of water at the time t and distance r away from the well axis; according to conservation of mass requirement, the following equation can be presented:

$$\nabla q = \frac{\partial C_S}{\partial t}.$$

(1)

Consider that

$$q = D_{eff} \nabla C_S,$$

(2)

Where ∇ is the gradient operator and D_{eff} is the coefficient of water diffusion.

According to the above equations, water diffusion equation can be established as follows:

$$\frac{\partial C_S}{\partial t} - \frac{1}{r}\frac{\partial}{\partial r}\left(r\frac{\partial C_S}{\partial r}\right)D_{eff} = 0. \tag{3}$$

And the boundary conditions are

$$t = 0, \quad r_w \leq r \leq \infty, \quad C_S = C_0,$$

$$t > 0, \quad r = r_w, \quad C_S = C_{df},$$

$$t > 0, \quad r \longrightarrow \infty, \quad C_S = C_0, \tag{4}$$

Where r_w is the wellbore radius; C_{df} is the saturated water content of shale; C_0 is the original water content. Sign $u = r/\sqrt{D_{eff}t}$, $(r, t) = \phi(u)$, then the following equations can obtain that

$$\frac{\partial C_S}{\partial t} = \frac{d\phi}{du}\cdot\frac{\partial u}{\partial t} = \frac{d\phi}{du}\left(-\frac{1}{2}u\right)t^{-1},$$

$$\frac{\partial C_S}{\partial r} = \frac{d\phi}{du}\cdot\frac{\partial u}{\partial r} = \frac{d\phi}{du}\frac{1}{\sqrt{D_{eff}t}},$$

$$\frac{\partial^2 C_S}{\partial r^2} = \frac{d}{dr}\left(\frac{d\phi}{du}\frac{1}{\sqrt{D_{eff}t}}\right) = \frac{1}{D_{eff}t}\frac{\partial^2 \phi}{\partial u^2} \tag{5}$$

Insert (5) to (3), then (6) can be obtained:

$$\phi'' = -\left(\frac{u}{2} + \frac{1}{u}\right)\phi'. \tag{6}$$

Equations (7) and (8) can obtain by integrating (6) that

$$\phi' = -Au^{-1}e^{-u^2/4},$$

(7)

$$C_S(r,t) = \phi(u) = \frac{A}{2}\int_{u^2/4}^{+\infty} x^{-1}e^{-x}dx + B.$$

(8)

Combining (8) and (4) the following equations can obtain that.

$$A = \frac{2\left(C_{df} - C_0\right)}{\int_{r_w^2/4D_{eff}t}^{+\infty} x^{-1}e^{-x}dx},$$

$$B = C_0.$$

(9)

Thus, the water content of shale formation around the wellbore can be written as follows:

$$C_S$$

$$= C_0 + \left(C_{df} - C_0\right)$$

$$\times\left[1 + \frac{2}{\pi}\int_0^\infty e^{-D_{eff}\zeta^2 \cdot t} \frac{J_0(\zeta r)\,N_0(\zeta r_w) - N_0(\zeta r)\,J_0(\zeta r_w)}{J_0^2(\zeta r_w) + N_0^2(\zeta r_w)}\right.$$

$$\left.\cdot\frac{d\zeta}{\zeta}\right],$$

(10)

Where $J_0()$ and $N_0()$ are the zero order of Bessel's functions of group one and two, respectively.

In a short period of time after drilling and within a short distance from the wellbore wall, (10) can be simplified to

$$C_S = C_0 + \left(C_{df} - C_0\right)\sqrt{\frac{r_w}{r}}\,\mathrm{erfc}\left(\frac{r - r_w}{2\sqrt{D_{eff}t}}\right).$$

(11)

The water diffusion character of shale is measured using this designed experiment equipment. All the shale core samples used in this paper were collected from Bohai Bay Basin of China. The drilling fluid which contacted the shale in this experiment was KCL drilling fluid. The experimental confining pressure was 20 MPa, and the differential pressure of the fluid was 6 MPa. Core samples were taken out after 50 hours and then cut into pieces to measure the water content of each piece. Three samples were tested in this research. The experimental results of core sample 1-1 are shown in Figure 2. Substituting the experimental results into (11), the coefficient of water diffusion of shale can be obtained. All the calculated water diffusion coefficients and clay mineral contents of these three samples are shown in Table 1. Smectite is the mineral most prone to hydration [59, 60], and the water diffusion coefficient is higher with more smectite.

Table 1: Clay mineral contents and water diffusion coefficient of shale

Core no.	Clay mineral total contents (%)	Clay mineral relative contents (%)				D_{eff} (cm²/h)
		Smectite	Illite	Kaolinite	Chlorite	
1-1#	21.68	45	19	17	19	0.0238
1-2#	29.28	51	24	13	12	0.0247
1-3#	26.68	32	20	26	22	0.0184

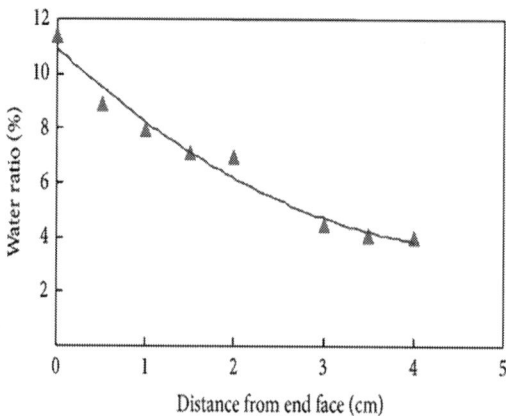

Figure 2: Experimental results of water diffusion in shale.

Chemical Effect of Drilling Fluid on Shale Mechanical Properties

The mechanical properties of shale can be altered seriously after contacting with drilling fluid. Existing forms of water in shale mainly include water vapor, solid water, bound water, adsorption water (film water), capillary water, and gravity water (free water) (Figure 3). Owing to the direct contact with drilling fluid around the wellbore, the free water of drilling fluid diffuses into shale under physical and chemical driving force. During the drilling process, the absorption water will increase, and the diffusion layer of rock particle will thicken, which will cause volume increase of shale and produce swelling stress. In order to calculate the swelling stress caused by hydration, the relation between water absorption and the swelling must be researched first through experiments. The experiment methods are similar to that used by Yew et al. [29]. The experimental results are shown in Figure 4.

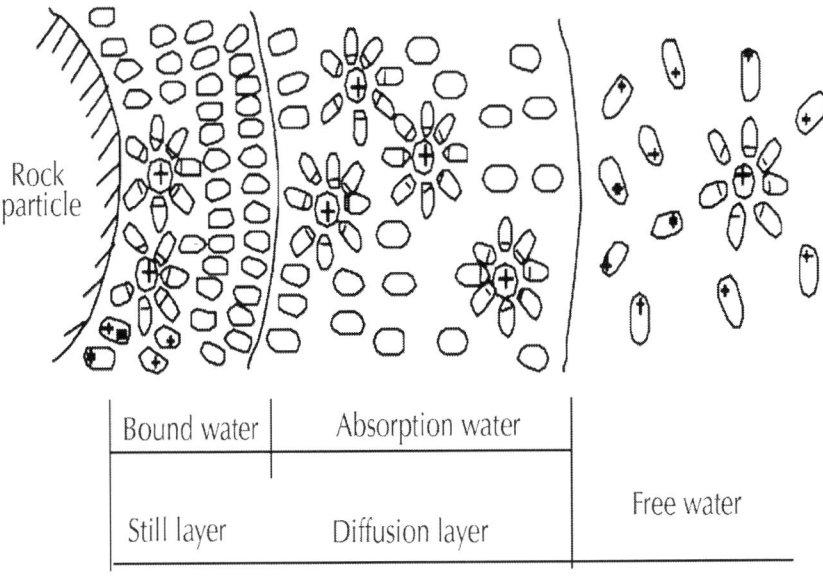

Figure 3: Water existing states in shale [63].

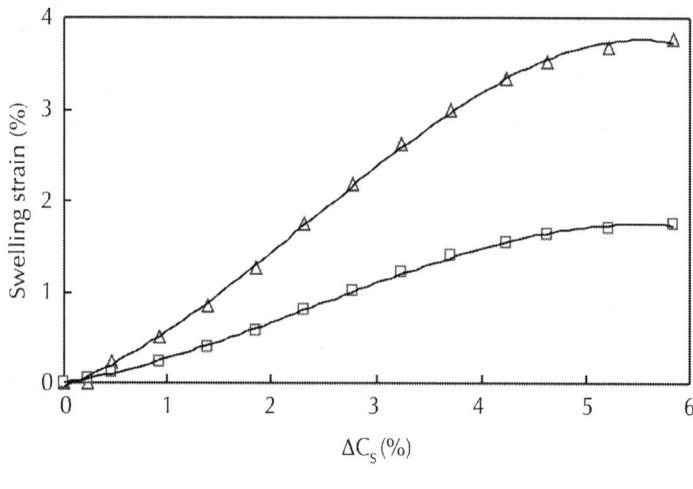

A Vertical strain

□ Horizontal strain

Figure 4: Experimental results of shale swelling.

The experiment results show that the swelling strain in the direction perpendicular to the deposition surface is larger than that of the parallel direction, which is resulted from the difference of drainage and stress conditions in different directions in sedimentation [61, 62]. The relationship of water content and swelling strain is as follows:

$$\varepsilon_v = -324.8(\Delta C_S)^3 + 23.5(\Delta C_S)^2 + 0.37\Delta C_S,$$

$$\varepsilon_h = -150.8(\Delta C_S)^3 + 10.9(\Delta C_S)^2 + 0.17\Delta C_S,$$

$$(12)$$

Where ε_v and ε_h are the swelling strain in the direction perpendicular and parallel to the deposition surface, respectively; C_s is the water content increment.

When contacting with drilling fluid, water sensitivity minerals of shale will absorb water and make a chemical reaction. Clay mineral in shale will react with the ions in drilling fluid [63, 64]:

$$K_{0.9}Al_{2.9}Si_{3.1}O_{10}(OH)_2 + nH_2O$$

$$\longrightarrow K_{0.9}Al_{2.9}Si_{3.1}O_{10}(OH)_2 nH_2O. \qquad (13)$$

The properties of shale particle will change by above chemical reaction, and then weaken the cohesive force between shale particles, which will soften the shale and weaken the strength.

In order to evaluate wellbore stability after shale hydration, the mechanical properties of shale after hydration must be researched. In order to ensure the uniformity of the core samples used in the experiment, the compressive acoustic wave velocities of core samples were tested. Only the samples whose velocity is close are chosen. The core samples are shown in Figure 5. The samples are immersed in KCL drilling fluid at the temperature of 55°C. MTS-816 rock test system (Figure 6) was adopted to test shale mechanical properties. The experiment results are showed in Table 2.

Table 2: Experiment results of cores immersing in drilling fluid

Immersing time (h)	0	2	6	12	24	48	96	192
Water content increment ΔC_s (%)	0	0.82	2.03	2.70	3.28	4.46	5.11	5.68
UCS (MPa)	16.30	13.56	11.76	12.24	9.48	8.28	8.52	6.60
Poisson's ratio	0.26	0.30	0.32	0.30	0.34	0.38	0.41	0.40
Elastic modulus (MPa)	5518.37	4055.17	2881.29	2176.90	2111.45	1759.96	2165.25	1599.73

Figure 5: Standard samples of shale.

Figure 6: MTS-816 rock test system.

The relationship of rock mechanical parameters with water content increment is obtained by the experimental results:

$$E = 5.5 \times 10^3 e^{-4.5\sqrt{\Delta C_S}},$$

$$v = 0.26 + 2.1\Delta C_S, \tag{14}$$

$$UCS = 16.3 - 1.49\Delta C_S, \tag{15}$$

Where E is the Young's modulus; v is the Poisson's ratio; UCS is the unconfined compressive strength.

Assume that the variation of shale cohesion with water content increment is the same with the variation of UCS:

$$\tau = \tau' - 1.49\Delta C_S, \tag{16}$$

Where τ is the cohesion; τ' is the initial cohesion before immersing, which is tested as 5.2 MPa.

The shear failure of wellbore obeys Mohr-Coulomb strength criterion. Mohr-Coulomb strength criterion can be expressed by principal stress [23]:

$$\sigma_1 = \sigma_3 \text{ctg}^2\left(45° - \frac{\varphi}{2}\right) + 2\tau \cdot \text{ctg}\left(45° - \frac{\varphi}{2}\right),$$

(17)

Where σ_1 and σ_3 are the maximum and minimum effective principal stresses, respectively; φ is the internal friction angle.

There is $\sigma_3 = 0$ in the uniaxial compression experiment. Inserting (15) and (16) into (17), the variation of internal friction angle with water content increment can be obtained:

$$\varphi = \arctan\left(\frac{16.3 - 1.49\Delta C_S}{10.4 - 2.98\Delta C_S}\right).$$

(18)

WELLBORE STABILITY MODEL WITH CHEMICAL-MECHANICAL COUPLING

Assuming that the shale is linear elastic material, the constitutive equation in plane strain is shown as follows:

$$\varepsilon_r = \frac{\left[\sigma_r - v\left(\sigma_\theta + \sigma_z\right)\right]}{E} + \varepsilon_h,$$

$$\varepsilon_\theta = \frac{\left[\sigma_\theta - v\left(\sigma_r + \sigma_z\right)\right]}{E} + \varepsilon_h,$$

$$\varepsilon_z = \frac{\left[\sigma_z - v\left(\sigma_r + \sigma_\theta\right)\right]}{E} + \varepsilon_v,$$

(19)

Where σ_r, σ_θ, and σ_z are radial, tangential, and vertical stresses, respectively; ε_r, ε_θ, and ε_z are radial, tangential, and vertical strains, respectively.

The formation is replaced by fluid column pressure after drilling. The original stress balance around the wellbore is broken. A new balance will be built. The stress balance equation is as follows [65]:

$$\frac{d\sigma_r}{dr} + \frac{\sigma_r - \sigma_\theta}{r} = 0,$$

(20)

Where σ_{ij} is the total stress tensor and f_i is the volume force.

The strain components and displacement components of the formation should meet the following geometric equation [65]:

$$\varepsilon_r = \frac{du}{dr},$$

$$\varepsilon_\theta = \frac{u}{r},$$

(21)

Where ε_{ij} is total strain tensor and u_i is the displacement component. Inserting (19) and (20) to (21), the following obtains that.

$$r\frac{d^2\sigma_r}{dr^2} + \left(3 - \frac{r}{E_1}\frac{dE_1}{dr} + \frac{2vr}{v^2-1}\frac{dv}{dr}\right)\frac{d\sigma_r}{dr}$$

$$+ \left(\frac{4v+1}{v^2-1}\frac{dv}{dr} - \frac{1}{E_1}\frac{2v-1}{v-1}\frac{dE_1}{dr}\right)\sigma_r$$

$$= \frac{E_1(1+v)}{v^2-1}\frac{d\varepsilon_v}{dr} + \frac{E_1\varepsilon_v}{v^2-1}\frac{dv}{dr}.$$

(22)

The boundary conditions for drilling are as follows [66]:

$$\sigma_r = P_w, \quad r = r_w,$$

$$\sigma_r = S, \quad r = \infty,$$

(23)

Where P_w is under the fluid column pressure and S is the far field horizontal in situ stress.

Solving (22) and (23) using finite-difference method, stress distribution around the wellbore and its change rules with drilling time are obtained. Combined with Mohr-Coulomb failure criterion, the time-dependent collapse pressure can be obtained.

TIME-DEPENDENT COLLAPSE PRESSURE

Based on the above model and experimental results, the variations of mechanical parameters of shale around the wellbore and time-dependent collapse pressure are analyzed. The calculation parameters are as follows: well depth $H = 1800$m, the initial water content $C_0 = 4\%$, saturation water content $C_{df} = 11.4\%$, the water diffusion coefficient $D_{eff} = 0.0238$ cm^2/h, and the wellbore radius $r_w = 10.8$ cm; the other parameters are obtained by the experimental results.

Figure 7 shows the variation of water content in shale formation around the wellbore with different open-hole time. The water content at the wellbore wall reaches to saturated state quickly after the wellbore is opened; in the same time, the water content would decrease with the increment of distance from the wellbore axis, and the decreasing rate is the highest near the wellbore wall. Thus, a hydrated area would develop around the wellbore. When the distance from the hole axis excesses 20 cm, the water content of shale almost no longer changes with the time increases, and the water content approaches the initial water content; in hydrated area, the longer the time, the more the shale water content when the distance is constant.

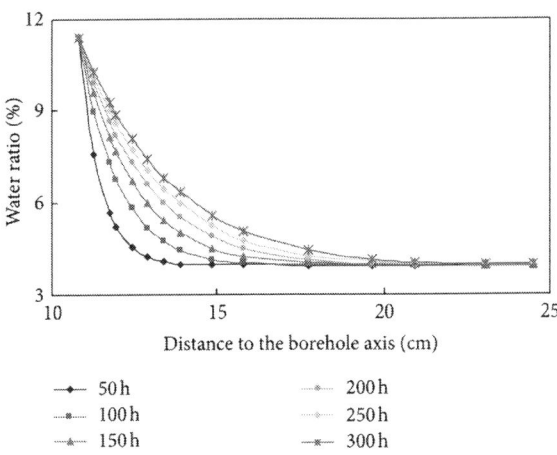

Figure 7: Water content distribution around the wellbore.

The distribution character of UCS of shale around the wellbore is presented in Figure 8. When a wellbore is opened, the UCS would decrease as the time increases. When the time is constant, the UCS would increase as the distance from the wellbore axis increases. The increasing rate near the wellbore wall is the highest.

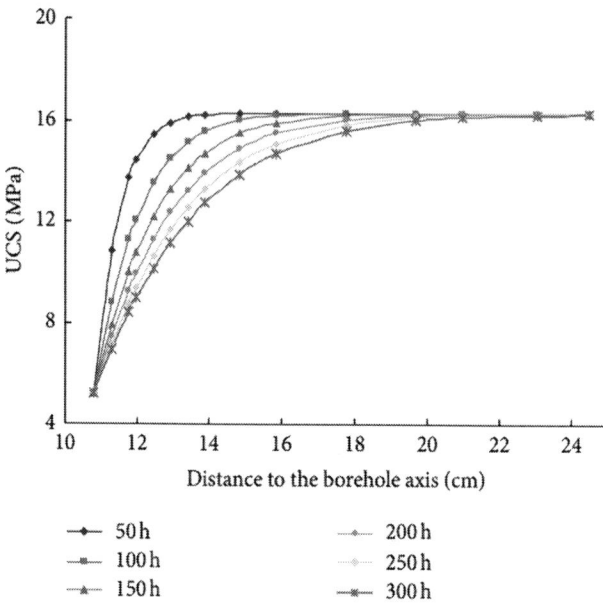

Figure 8: UCS distribution around the wellbore.

The variation of collapse pressure with drilling time under different water diffusion coefficients is shown in Figure 9. The results show that the collapse pressure increases rapidly in a short time after the wellbore opening due to shale hydration. Then the increasing rate of collapse pressure would decrease. At last, the collapse pressure would increase linearly with a very low increasing rate. After the wellbore is opened 10 days, the collapse pressure nearly no longer changes. According to Figure 9, after the wellbore opening, increasing the drilling fluid density to 1.45 g/cm³ gradually is more beneficial for long-time wellbore stability when the water diffusion coefficient D_{eff}=0.0238cm²/h. The higher the water diffusion coefficient is, the larger the increasing range of collapse is. On the other hand, the possibility of wellbore instability will be increasing with more smectite in shale.

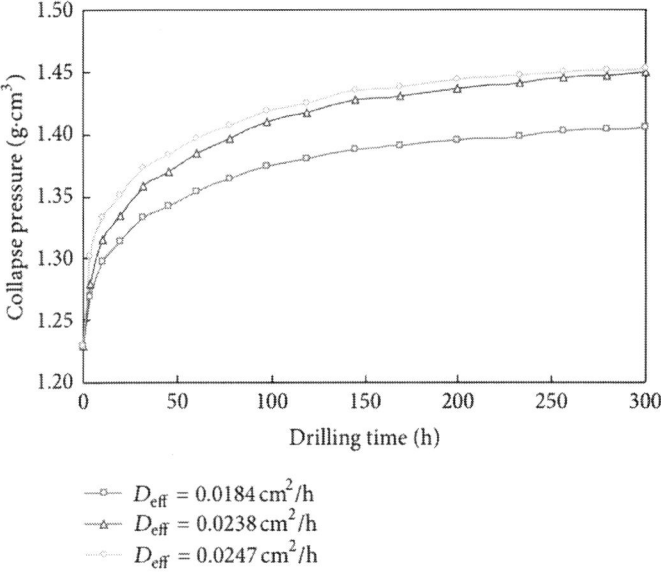

Figure 9: Time-dependent collapse pressure.

CONCLUSIONS

Water content at different distance from the end face of shale sample is measured using the designed equipment in the condition of down whole temperature and pressure.

The water content of shale at the wellbore wall reaches to saturated state quickly when the wellbore is opened; the water content of shale would decrease with the increment of distance from wellbore axis, and the decreasing rate is the highest near the wellbore wall.

Due to the impact of shale hydration, the strength of the circumferential formation around the well is gradually reduced with the increase of drilling time and increases with the increase of the distance away from the wellbore.

Collapse pressure of shale increases sharply in a short time after drilling and then slows down. The collapse pressure is basically steady after several days of the open-hole time. The initial stable wellbore may collapse with the increase of the open-hole time.

Shale containing more smectite is more prone to react with drilling fluid. The possibility of wellbore instability of shale is higher with more smectite as the increasing range of collapse pressure is larger.

ACKNOWLEDGMENTS

This work is financially supported by Science Fund for Creative Research Groups of the National Natural Science Foundation of China (Grant no. 51221003), National Natural Science Foundation Project of China (Grant no. 51134004 and Grant no. 51174219), and National Oil and Gas Major Project of China (Grant no. 2011ZX05009-005 and Grant no. 2011ZX05026-001-01).

REFERENCES

1.   W. B. Bradley, "Mathematical concept-stress cloud can predict borehole failure," Oil & Gas Journal, vol. 77, no. 8, pp. 92–101, 1979.

2.   F. J. Santarelli, E. T. Brown, and V. Maury, "Analysis of Borehole stresses using pressure-dependent, linear elasticity," International Journal of Rock Mechanics and Mining Sciences and, vol. 23, no. 6, pp. 445–449, 1986.

3.   B. S. Aadnoy, "Introduction to special issue on borehole stability," Journal of Petroleum Science and Engineering, vol. 38, no. 3-4, pp. 79–82, 2003. ·

4.   B. S. Aadnøy and M. Belayneh, "Elasto-plastic fracturing model for wellbore stability using non-penetrating fluids," Journal of Petroleum Science and Engineering, vol. 45, no. 3-4, pp. 179–192, 2004. · ·

5.   L. Bailey, J. H. Denis, and G. C. Maitland, "Drilling fluids and wellbore stability current performance and future challenges," in Chemicals in the Oil Industry, P. H. Ogden, Ed., Royale Society of Chemistry, London, UK, 1991.

6.   J. S. Bell, "Practical methods for estimating in situ stresses for borehole stability applications in sedimentary basins," Journal of Petroleum Science and Engineering, vol. 38, no. 3-4, pp. 111–119, 2003. · ·

7. Y. Wang and M. B. Dusseault, "A coupled conductive-convective thermo-poroelastic solution and implications for wellbore stability," Journal of Petroleum Science and Engineering, vol. 38, no. 3-4, pp. 187–198, 2003. · ·

8. M. D. Zoback, C. A. Barton, M. Brudy et al., "Determination of stress orientation and magnitude in deep wells," International Journal of Rock Mechanics and Mining Sciences, vol. 40, no. 7-8, pp. 1049–1076, 2003. · ·

9. A. M. Al-Ajmi and R. W. Zimmerman, "Stability analysis of vertical boreholes using the Mogi-Coulomb failure criterion," International Journal of Rock Mechanics and Mining Sciences, vol. 43, no. 8, pp. 1200–1211, 2006. · ·

10. T. Al-Bazali, J. Zhang, M. E. Chenevert, and M. M. Sharma, "Factors controlling the compressive strength and acoustic properties of shales when interacting with water-based fluids," International Journal of Rock Mechanics and Mining Sciences, vol. 45, no. 5, pp. 729–738, 2008. · ·

11. M. E. Zeynali, "Mechanical and physico-chemical aspects of wellbore stability during drilling operations," Journal of Petroleum Science and Engineering, vol. 82-83, pp. 120–124, 2012. · ·

12. J. Zhang, J. Lang, and W. Standifird, "Stress, porosity, and failure-dependent compressional and shear velocity ratio and its application to wellbore stability," Journal of Petroleum Science and Engineering, vol. 69, no. 3-4, pp. 193–202, 2009.

13. F. K. Mody and A. H. Hale, "Borehole-stability model to couple the mechanics and chemistry of drilling-fluid/shale interactions," Journal of Petroleum Technology, vol. 45, no. 11, pp. 1093–1101, 1993.

14. G. Chen, M. E. Chenevert, M. M. Sharma, and M. Yu, "A study of wellbore stability in shales including poroelastic, chemical, and thermal effects," Journal of Petroleum Science and Engineering, vol. 38, no. 3-4, pp. 167–176, 2003. · ·

15. L. C. Coelho, A. C. Soares, N. F. F. Ebecken, J. L. D. Alves, and L. Landau, "The impact of constitutive modeling of porous rocks on 2-D wellbore stability analysis," Journal of Petroleum Science and Engineering, vol. 46, no. 1-2, pp. 81–100, 2005.

16. H. C. H. Darley, "A laboratory investigation of borehole stability," Journal of Petroleum Technology, vol. 246, pp. 821–826, 1969.

17. R. T. Ewy and N. G. W. Cook, "Deformation and fracture around cylindrical openings in rock-I. Observations and analysis of deformations," International Journal of Rock Mechanics and Mining Sciences and, vol. 27, no. 5, pp. 387–407, 1990.

18. O. A. Helstrup, Z. Chen, and S. S. Rahman, "Time-dependent wellbore instability and ballooning in naturally fractured formations," Journal of Petroleum Science and Engineering, vol. 43, no. 1-2, pp. 113–128, 2004. · ·

19. J. L. Yuan, J. G. Deng, Q. Tan, B. H. Yu, and X. C. Jin, "Borehole stability analysis of horizontal drilling in shale gas reservoirs," Rock Mechanics and Rock Engineering. In press.

20. J. C. Roegiers, "Well modeling: an overview," Oil and Gas Science and Technology, vol. 57, no. 5, pp. 569–577, 2002.

21. M. D. Zoback, D. Moos, L. Mastin, and R. N. Anderson, "Well bore breakouts and in situ stress,"Journal of Geophysical Research, vol. 90, no. 7, pp. 5523–5530, 1985.

22. C. A. Barton, M. D. Zoback, and K. L. Burns, "In-situ stress orientation and magnitude at the Fenton Geothermal site, New Mexico, determined from wellbore breakouts," Geophysical Research Letters, vol. 15, no. 5, pp. 467–470, 1988.

23. E. Fjær, R. M. Holt, P. Horsrud, et al., Petroleum Related Rock Mechanics, Elsevier, 2nd edition, 2008.

24. R. Narayanasamy, D. Barr, and A. Milne, "Wellbore instability predictions within the cretaceous mudstones, clair field, west of shetlands," in Offshore Europe, Aberdeen, 2009.

25. E. T. Brown, J. W. Bray, and F. J. Santarelli, "Influence of stress-dependent elastic moduli on stresses and strains around axisymmetric boreholes," Rock Mechanics and Rock Engineering, vol. 22, no. 3, pp. 189–203, 1989. · ·

26. P. A. Nawrocki, M. B. Dusseault, and R. K. Bratli, "Assessment of some semi-analytical models for non-linear modeling of borehole stresses," International Journal of Rock Mechanical & Mining Science, vol. 35, no. 4-5, pp. 522–531, 2002.

27. V. M. Maury and J. M. Sauzay, Borehole Instability: Case Histories, Rock Mechanics Approach, and Results, SPE, 1987.

28. J. C. Roegiers, "Well modeling: an overview," Oil and Gas Science and Technology, vol. 57, no. 5, pp. 569–577, 2002.

29. C. H. Yew, M. E. Chenevert, E. Martin, et al., "Wellbore stress distribution produced by moisture adsorption," SPE Drilling Engineering, vol. 5, no. 4, pp. 311–316, 1990.

30. E. van Oort, "On the physical and chemical stability of shales," Journal of Petroleum Science and Engineering, vol. 38, no. 3-4, pp. 213–235, 2003. · ·

31. Z. Qiu, J. Xu, K. Lu, L. Yu, W. Huang, and Z. Wang, "Multivariate cooperation principle for well-bore stabilization," Acta Petrolei Sinica, vol. 28, no. 2, pp. 117–119, 2007.

32. M. Yu, M. E. Chenevert, and M. M. Sharma, "Chemical-mechanical wellbore instability model for shales: accounting for solute diffusion," Journal of Petroleum Science and Engineering, vol. 38, no. 3-4, pp. 131–143, 2003. · ·

33. A. Ghassemi and A. Diek, "Linear chemo-poroelasticity for swelling shales: theory and application,"Journal of Petroleum Science and Engineering, vol. 38, no. 3-4, pp. 199–212, 2003. · ·

34. B. S. Aadnoy, "A complete elastic model for fluid-induced and in-situ generated stresses with the presence of a borehole," Energy Sources, vol. 9, no. 4, pp. 239–259, 1987.

35. L. Cui, Y. Abousleiman, A. H.-D. Cheng, and J.-C. Roegiers, "Time-dependent failure analysis of inclined boreholes in fluid-saturated formations," Journal of Energy Resources Technology, vol. 121, no. 1, pp. 31–39, 1999.

36. M. E. Chenevert and V. Pernot, "Control of shale swelling pressures using inhibitive water-base muds," in Proceedings of the 67th SPE Annual Technical Conference and Exhibition, pp. 27–30, SPE, New Orleans, La, USA, 1998.

37. P. T. Chee and G. R. Brian, Effects of Swelling and Hydration Stress in Shale on Wellbore Stability, SPE, 1997.

38. P. T. Chee and G. R. Brian, "Integrated rock mechanics and drilling fluid design approach to manage shale instability," in SPE/ISRM Rock Mechanics in Petroleum Engineering, 1998. ·

39. J. P. Simpson, H. L. Dearing, and D. P. Salisbury, "Downhole simulation cell shows unexpected effects on shale hydration on borehole wall," SPE Drilling Engineering, vol. 4, no. 1, pp. 24–30, 1989.

40. A. H. Hale, F. K. Mody, and D. P. Salisbury, "Experimental investigation of the influence of chemical potential on wellbore stability," in Drilling Conference, pp. 377–389, February 1992.

41. F. K. Mody and A. H. Hale, "Borehole stability model to couple the mechanics and chemistry of drilling fluid shale interaction," in Proceedings of the SPE/IADC Drilling Conference, pp. 473–490, February 1993.

42. E. Oort, A. H. Hale, and F. K. Mody, "Manipulation of coupled osmotic flows for stabilisation of shales exposed to water-based drilling fluids," in Proceedings of the SPE Annual Technical Conference and Exhibition, pp. 497–509, October 1995.

43. C. P. Tan, M. Amanullah, F. K. Mody, and U. A. Tare, "Novel high membrane efficiency water-based drilling fluids for alleviating problems in troublesome shale formations," in Proceedings of the IADC/SPE Asia Pacific Drilling Technology, pp. 63–72, November 2002.

44. M. E. Chenevert, "Shale Alteration by Water Adsorption," Journal of Petroleum Technology, vol. 22, no. 9, pp. 1141–1148, 1970.

45. G. Z. Chen, A study of wellbore stability in shales including poroelastic, chemical, and thermal effects [Ph.D. dissertation], The University of Texas at Austin, 2001.

46. Y. H. Lu, M. Chen, Y. Jin, X. Q. Teng, W. Wu, and X. Q. Liu, "Experimental study of strength properties of deep mudstone under drilling fluid soaking," Chinese Journal of Rock Mechanics and Engineering, vol. 31, no. 7, pp. 1399–1405, 2012.

47. R. Z. Huang, M. Chen, and J. G. Deng, "Study on shale stability of wellbore by mechanics coupling with chemistry method," Drilling Fluid & Completion Fluid, vol. 12, no. 3, pp. 15–21, 1995.

48. A. H. Hale and F. K. Mody, "Borehole-stability model to couple the mechanics and chemistry of drilling-fluid/shale interactions," Journal of Petroleum Technology, vol. 45, no. 11, pp. 1093–1101, 1993.

49. A. H. Hale, F. K. Mody, and D. P. Salisbury, "Influence of chemical potential on wellbore stability," SPE Drilling and Completion, vol. 8, no. 3, pp. 207–216, 1993.

50. J. Deng, D. Guo, J. Zhou, and S. Liu, "Mechanics-chemistry coupling calculation model of borehole stress in shale formation and its numerical solving method," Chinese Journal of Rock Mechanics and Engineering, vol. 22, no. 1, pp. 2250–2253, 2003.

51. L. W. Zhang, D. H. Qiu, and Y. F. Cheng, "Research on the wellbore stability model coupled mechanics and chemistry," Journal of Shandong University, Engineering Science, vol. 39, no. 3, pp. 111–114, 2009.

52. A. Ghassemi, Q. Tao, and A. Diek, "Influence of coupled chemo-poro-thermoelastic processes on pore pressure and stress distributions around a wellbore in swelling shale," Journal of Petroleum Science and Engineering, vol. 67, no. 1-2, pp. 57–64, 2009. · ·

53. Q. Wang, Y. Zhou, Y. Tang, and Z. Jiang, "Analysis of effect factor in shale wellbore stability," Chinese Journal of Rock Mechanics and Engineering, vol. 31, no. 1, pp. 171–179, 2012.

54. Q. Wang, Y. C. Zhou, G. Wang, H. W. Jiang, and Y. S. Liu, "A fluid-solid-chemistry coupling model for shale wellbore stability," Petroleum Exploration and Development, vol. 39, no. 4, pp. 475–480, 2012.

55. M. E. Chenevert and A. K. Sharma, "Permeability and effective pore pressure of shales," SPE Drilling & Completion, vol. 8, no. 1, pp. 28–34, 1993.

56. V. X. Nguyen, Y. N. Abousleiman, and S. K. Hoang, "Analyses of wellbore instability in drilling through chemically active fractured-rock formations," SPE Journal, vol. 14, no. 2, pp. 283–301, 2009.

57. E. van Oort, "Physico-chemical stabilization of shales," in Proceedings of the SPE International Symposium on Oilfield Chemistry, pp. 523–538, February 1997.

58. E. van Oort, "On the physical and chemical stability of shales," Journal of Petroleum Science and Engineering, vol. 38, no. 3-4, pp. 213–235, 2003. · ·

59. E. Oort, A. H. Hale, F. K. Mody, and S. Roy, "Critical parameters in modelling the chemical aspects of borehole stability in shales

and in designing improved water-based shale drilling fluids," inProceedings of the SPE Annual Technical Conference & Exhibition, pp. 171–186, September 1994.

60. M. Chen, Y. Jin, and G. Q. Zhang, Petroleum Engineering Related Rock Mechanics, Science Press, Beijing, China, 2008.

61. C. E. Weaver, Clays, Muds, and Shales, Elsevier, 1989.

62. S. H. Ong, Borehole Stability [Ph.D. dissertation], The U. Of Oklahoma, 1994.

63. X. J. Zhu, "Water-weakening properties of soften rocks," Technology of Mineral Science, vol. 3-4, pp. 46–50, 1996.

64. C.-H. Yang, H.-J. Mao, X.-C. Wang, X.-H. Li, and J.-W. Chen, "Study on variation of microstructure and mechanical properties of water-weakening slates," Rock and Soil Mechanics, vol. 27, no. 12, pp. 2090–2098, 2006.

65. Z. L. Xu, Elastic Mechanics, Higher Education Press, Beijing, China, 1998.

66. J. G. Deng, "Calculation method of mud density to control borehole closure rate," Chinese Journal of Rock Mechanics and Engineering, vol. 16, no. 6, pp. 522–528, 1997.

Chapter **4**

Observation of Drilling Burr and Finding out the Condition for Minimum Burr Formation

Nripen Mondal, Biswajit Sing Sardar, Ranendra Nath Halder, and Santanu Das

Department of Mechanical Engineering, Kalyani Government Engineering College, Kalyani, West Bengal 741235, India

ABSTRACT

Suppression or elimination of burr formation at the exit edge of the workpiece during drilling is essential to make quality products. In this work, low alloy steel specimens have been drilled to observe burr height under different machining conditions. Taper shank, uncoated 14 mm diameter HSS twist drills are used in these experiments. Dry environment is maintained in experiment set I. Water is applied as cutting fluid in experiment set II. In the next four sets of experiments, the effect of providing back-up support material and exit edge bevel is observed on formation of burr at the exit edge of specimens under dry and wet conditions. It is revealed that an exit edge bevel of 31 degrees with water as the cutting fluid gives negligible burr at the exit edge of the drilled hole at certain machining conditions. Use of a back-up support can also reduce drill burr to a large extent. In this paper, artificial neural networks (ANN) are developed for modeling experimental results, and modeled values show close matching with the experimental results with small deviations.

INTRODUCTION

Drilling is a common hole-making operation, and the majority of workpieces are subject to hole-drilling before they leave a machine shop. However, presence of burr on the drilled workpieces creates problems not only in handling, but also in the assembly line. Burrs are undesired projections attached to the edge of drilled holes. These are found to be substantially greater at exit side than entry side. Hence, elimination, or large-scale reduction, of exit burr is the necessary requirement of an industry [1–5].

Many researchers worked on burr-related problems associated with drilling and other processes and also investigated the mechanism behind burr formation. Control charts were applied by Min [6] and Kim et al. [7] for control and prediction of burr height during drilling different steels. The same technique was also employed by Lee and Dornfeld [8] for estimating burr size during microdrilling to some success. Finite element method (FEM) was used by some others [9–11] to observe stress and deformation patterns analytically to understand the reason behind burr formation. Guo and Dornfeld [9] made finite element analysis to assess the substantial reduction of drilling burr with back-up support. They also successfully did [10] this analysis to understand burr formation in stainless steels. FEM was also used by Park and Dornfeld [11] to find out the influence of exit edge angle of a specimen, tool rake angle, and back-up support on burr formation. The estimates made showed remarkable results.

An overview on different strategies to control burr was presented by some researchers [1–5, 12, 13]. Investigations on the effect of drill size [14, 15], use of drills with special geometry [16–19], effect of using different coolants, exit edge modification, and providing back-up support [20–22] were explored by other research groups. Lee and Dornfeld [8] carried out experiments on four different materials using HSS and carbide drills of varying shapes and geometry. They concluded that step drills with less step angle and step size produced quite small burr. Min et al. [20] performed experimental and analytical investigation on drilling burr formation by varying interaction angle at the exit edge. It is the angle by which work exit surface is inclined. They observed that a large interaction angle results in quite less burr due to less bending of job material to form a burr. Another group of researchers found out [19] variable feed drilling with a suitable amplitude to give high tool life with quite low burr height. Although variable feed was reported to reduce burr remarkably compared to that of constant feed drilling, implementation of this method needs special facility to be provided on a drilling machine.

During tool/cutter exit from the workpiece, cutting edge of the tool/cutter was observed [5–12] to have been chipped off or broken beside formation of burr/foot. This problem also was seen to reduce with the introduction of special tool geometry, suitable exit order sequence, and work edge bevel. Provision of a beveled work edge was reported to lower burr height noticeably in milling and drilling operations [12–15, 21–23] due to gradually decreasing depth of cut and the reducing need of back-up support. Suitable tool path selection also reduces burr height [24]. Significant effect of different size of drills on burr formation was observed in some other works [14, 15] under varying cutting conditions. Introduction to different shape and geometry of cutting tool was also reported [8, 17, 18, 25] to reduce burr significantly for specific applications. A typical stepped drill was tested to restrict or to remove burr effectively. For laminated composites, delamination and burr formation at interface layers were studied [26] and some conditions were reported to reduce formation of burr and to avoid delamination.

In another work, mechanism of exit burr formation during drilling of aluminium alloys was closely studied [27], and an analytical model was proposed considering the effect of temperature that matched well with experimental results. Burr size was also modeled by other researchers [28] analytically for ductile metals to make good prediction of burr formation. The nature of burr formed during drilling of small size drills under dry and minimum quantity lubrication (MQL) with water soluble oil condition was discussed [29]. They also tried to obtain good quality holes with minimum burr by using optimized tool selection. Investigation of drill burr during hole-making on low carbon steel specimens with 40 Brinell Hardness Number was carried out by Roy et al. [23]. Utility of providing a back-up support or beveling the exit edge by 31° was observed in that work. However, effect of water-cooled condition was not explored suitably with or without a bevel in this work. Providing bevel in dry condition was reported not to show remarkable reduction in burr height.

Karnik et al. [30] tried to minimize burr size in drilling stainless steel workpieces with the use of genetic algorithm (GA) and Taguchi method of designing experiments. An innovative measurement technique to find out burr height was also reported [31]. Gaitonde and Karnik [32] utilized artificial neural networks (ANN) and particle swarm optimisation (PSO) approaches for optimal selection of drilling parameters to achieve minimum burr size (height and thickness) during drilling within the domain of experiments.

Even as a lot of works have been carried out on burr, still a lot of scope remains to find out appropriate strategy to minimize burr for a particular process of machining different workpiece materials. The obbjective of the present investigation is to explore effects of different machining conditions and strategies on drilling burr formation and selection of the optimal condition to suppress burr formation significantly. 14 mm diameter holes are drilled with or without using a back-up support and bevel at the exit edge to notice formation of burr in low alloy steel specimens under dry and wet with water cooled environment.

EXPERIMENTAL SETUP AND MACHINING CONDITION

Experiments are conducted in this work on a radial drilling machine (Make-Energy Limited, India) under dry and wet (water cooled) conditions. Low alloy steel specimens are taken for drilling experiments using 14 mm diameter drills.

Experimental conditions are given in Table 1. Cutting velocity, V_c, and feed, S_z, are chosen within 20–31 m/min and 0.032– 0.08 mm/rev, respectively. Ranges of V_c and S_z are selected considering usual industrial practice for drilling low alloy steels with tungsten grade HSS drill bit. Back-up support is provided to explore its influence on reduction of burr formation in some sets of experiments. It is done by placing similar specimens as that of the test piece with slightly less width below the test piece as shown in Figure 1 and without providing any edge bevel. Less width of the support facilitates easy clamping of test piece in a machine vice. Drilling is continued up to the middle of back-up support thickness.

Table 1: Experimental conditions

Machine tool	Radial drilling machine, Make: Energy Limited, India, Model: RDH-32/930
	Main motor power: 1.5 kW
Tool holder	R/L 265 ME-20 AL, Make: Sandvik Asia Limited, India
Cutting tool	Taper shank, uncoated 14 mm diameter HSS twist drill, Make: Miranda (India)
Workpiece material	Low alloy steel, hardness: 225 HB
	Composition: C (0.17%), Si (0.21%), Mn (0.63%), P (0.09%)
Workpiece size	100 mm × 50 mm × 6 mm
Cutting velocity	20, 25 and 31 m/min
Feed	0.032, 0.05 and 0.08 mm/rev
Environment	Dry and water-cooled
Experiment sets	I—Drilling in dry condition without any back-up support or edge bevel
	II—Drilling with water cooling without any back-up support or edge bevel
	III—Drilling in dry condition with a back-up support
	IV—Drilling in water-cooled condition with a back-up support
	V—Drilling in dry condition with 31° exit edge bevel
	VI—Drilling in water-cooled condition with 31° exit edge bevel

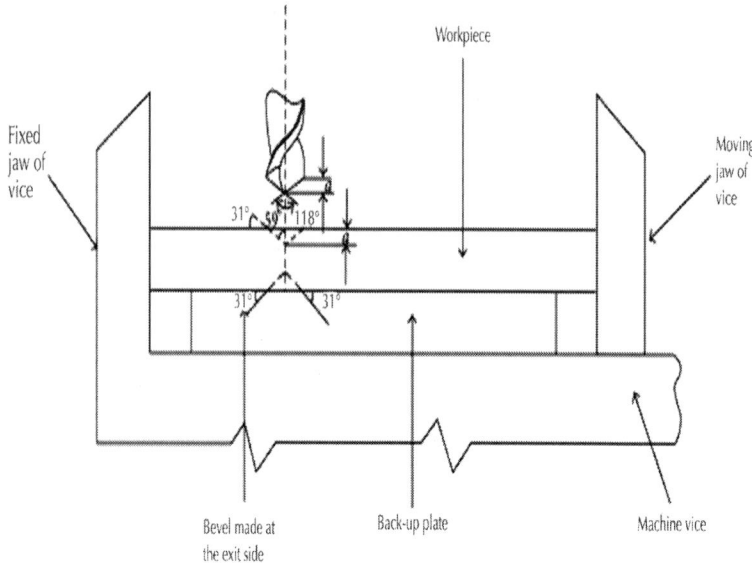

Figure 1: Schematic representation of drilling using a back-up plate and/or exit edge bevel.

Back-up plate is used to provide the required support during the tool exit and intends not to allow rotation of shear plane to a negative shear plane and suppresses bending of chip to form a burr. The shear plane rotation was reported to result in burr formation in a number of works [4, 5, 9, 13, 21]. Possible rotation of a positive shear plane to a negative one is shown schematically in Figure 2. When the tool or cutter approaches the emerging edge of a workpiece, no back-up material exits at that position to support the cutting forces exerted by the cutting tool onto the part of the workpiece. As a result, the shear plane gets oriented to the negative direction. If a back-up material is placed below the workpiece, then required support is likely to be provided by it resulting in expected reduction in the formation of burr. The effectiveness of suppressing formation of burr by using a back-up support is compared to that of using edge beveling in this work.

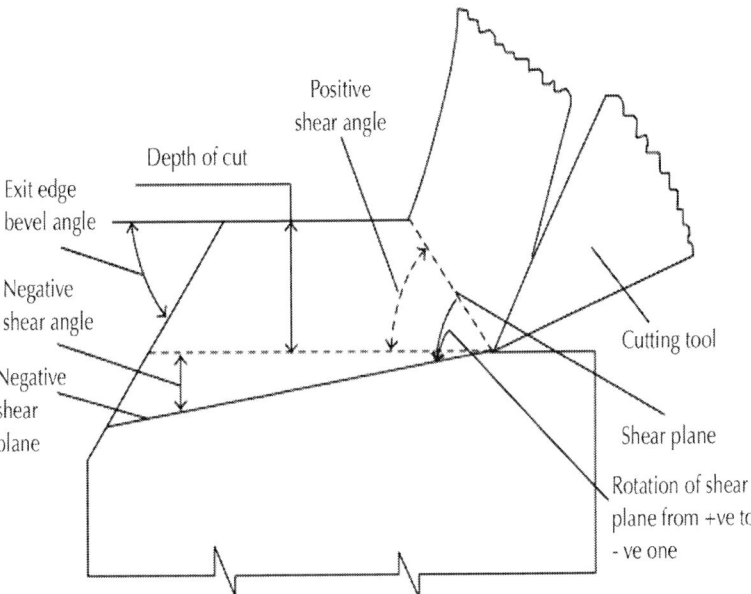

Figure 2: Schematic diagram showing orientation of positive shear plane to a negative one and exit edge bevel.

A bevel of 31° is provided at the rear side of the workpiece as shown in Figure 1 for experiment sets V and VI. The bevel is made with the help of drilling with the same 14 mm drill bit up to the drill point length, a, as shown in Figure 1. In this way, 31° exit edge bevel angle is made, as it is complementary to 59° which is half the point angle, 118°. Thus, no other processes are needed to produce the bevel. Only appropriate marking at the location of the bevel making with the drill is to be done. Bevel portion of the specimen is placed at the rear side of the workpiece aligning with the hole to make with suitable marking. When this exit edge bevel is used, no back-up support is provided.

Number of other parameters, such as lip clearance angle and point angle, may have an influence on burr formation, and this may be explored experimentally. However, investigation is restricted in this work to find out the effect on burr height by varying cutting velocity, feed, machining environment, and with and without the use of back-up and exit edge bevel of 31°.

Along the bevel provided at the exit edge of the workpiece, depth of cut gets gradually reduced. When the drill bit goes down along the bevel position, chip width gets reduced gradually. Correspondingly, the influence of chisel edge to deliver axial thrust causing bulging out of the thinning work material at the exit end is eliminated. When peripheral point of cutting edge of the drill bit approaches the rear end of the specimen, depth of cut reduces gradually needing correspondingly decreasing cutting force (thrust) and torque. Hence, at some bevel angle, it is expected that no substantial requirement of back-up support will be there leading to suppression of burr formation at the exit edge [4, 5, 7, 13, 20–22]. In the present work, 31° exit edge bevel is provided as it is easily achievable by using the twist drill point portion having a point angle of 118° as detailed in Figure 1. Influence of this is explored on the extent of suppressing burr formation.

Table 1 shows the detail of experimental conditions. Process parameters are selected in-line with that normally practiced in industry using standard HSS twist drills. Three levels of feed (S_z) and cutting velocity (V_c) are employed following full factorial design of experiments at each experiment set as detailed in Table 2. Hence, at each experiment set, $3^2 = 9$ tests are performed. Although a burr size may be characterized by its thickness and height, in the present work, burr height is considered to characterize a burr in line with many other works reported earlier [7, 18, 19, 27]. Burr height is measured using a vernier height gauge. Six different sets of experiments are conducted as detailed in Table 1 to find out the condition giving minimum drilling burr within the domain of experiments. Experiment sets I and II are performed under dry and wet with water cooling condition, respectively, without using any backing plate or edge bevel. Experiments with the provision of backing support are done under dry and water-cooled conditions in experiment sets III and IV, respectively. In experiment sets V and VI, drilling is carried out at respective dry and wet environments with an exit edge bevel of 31°.

Table 2: Test conditions in each experiment set

Sl. no.	Feed (SZ) (mm/rev)	Cutting velocity (Vc) (m/min)
1	0.032	20
2	0.032	25
3	0.032	31
4	0.050	20
5	0.050	25
6	0.050	31
7	0.080	20
8	0.080	25
9	0.080	31

RESULTS AND DISCUSSION

Discussion on the Observation of Drilling Burr at Dry Condition without Any Back-Up Support or Beveled Edge (Experiment Set I)

Observation on experiments in drilling at dry condition without using any back-up support material, or edge beveling, is presented in Figure 3. Large burrs are found in these tests. Variation of burr height is noticed at different machining conditions, although no clear trend is seen with the individual machining parameters. Large size burrs are expectedly formed in these tests as no back-up support material is available at the exit edge of the drilled hole. For this, shear plane is likely to have oriented to a negative direction facilitating formation of burr as shown in Figure 2 [21, 24]. When shear plane is oriented towards a negative plane, the chip is bent downwards and can be attached to the exit edge of workpiece forming a burr. Large extent of plastic deformation under dry condition with high temperature rise may have resulted in large burrs. Corresponding photographs of burrs observed are shown in Figure 4. For experiment set I, only at a cutting velocity V_c of 20 m/min and a feed S_z of 0.08 mm/rev (test No. 7), burr height of slightly less than 2 mm is found. In all the nine tests in experiment set I, transient, nonuniform burrs are observed. Classification of drilling burr was discussed by Kim et al. [7]. They classified drilling burr as uniform, crown, transient, and uniform with drill cap.

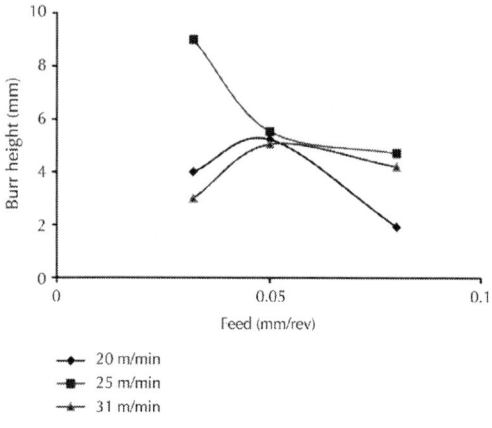

Figure 3: Plot of variation of burr height at different cutting velocity and feed for experiment set I.

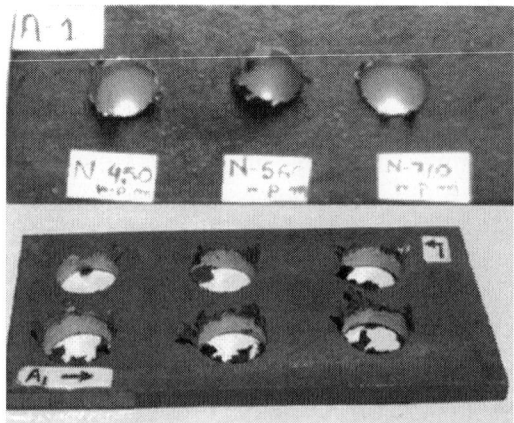

Figure 4: Photographic view of burr formed in experiment set I (left-right order: top for Sl. numbers 1–3, bottom for Sl. numbers 4–6, middle for Sl. numbers 7–9).

Discussion on the Observation of Drilling Burr under Water-Cooled Condition without Back-Up Support or Edge Bevel (Experiment Set II)

Experiment set II is conducted in water cooled condition without using a back-up material and edge bevel. The experimental results are plotted in Figure 5 showing the variation of burr height observed under different cutting conditions. No remarkable reduction in burr height is found to be there at water-cooled condition compared to that at dry environment. At certain machining conditions, burr height is lesser with water cooling than dry condition. However, at other conditions, larger burr height is observed with water cooled condition than that with dry condition. Figure 6 shows typical burrs seen at the exit edge of the drilled holes. In this experiment set II, mainly transient burrs are observed. Water cooling is expected to reduce drilling temperature and thereby may reduce the extent of plastic deformation during drilling. However, marginal contribution of water in lubricating tool-chip-workpiece interface regions may be the possible reason behind having no significant effect of it on burr reduction. In this experiment set also, no definite trend of variation of burr height with machining parameters is noticed as that of experiment set I.

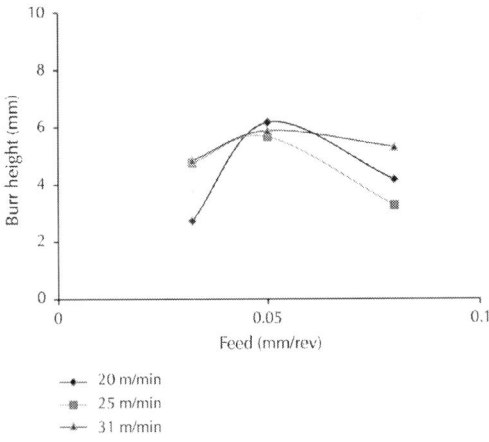

Figure 5: Plot of variation of burr height at different cutting velocity and feeds for experiment set II.

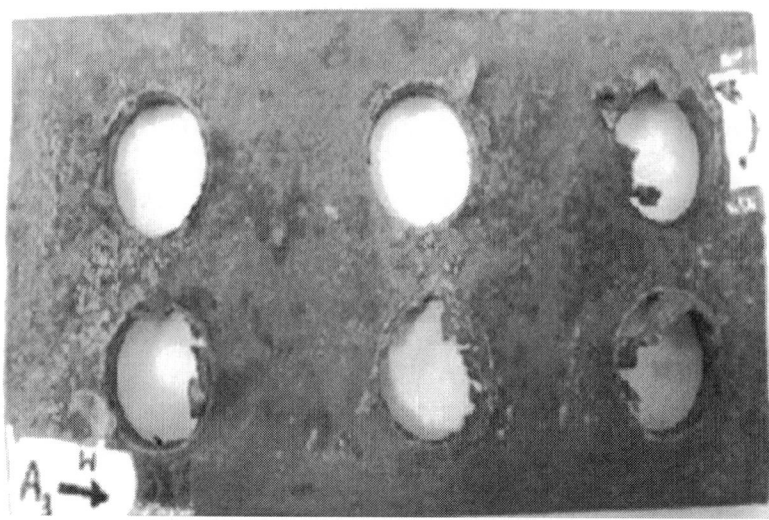

Figure 6: Photographic view of typical burr formed in experiment set II (left-right order: bottom for Sl. numbers 1–3, top for Sl. numbers 4–6).

Discussion on Drilling Burr Formation with a Back-Up Support at Dry Condition (Experiment Set III)

Experiment set III is performed in dry condition by providing a low alloy steel back-up support. This backing plate is similar to that of the test specimen with slightly less width to facilitate easy clamping of the test piece in a machine vice (Figure 1). This is done to render a support during tool exit, so that rotation of shear plane from positive value to a negative one about a pivot point is suppressed. With this, burr formation is expected to be restricted. Burrs formed at these conditions are observed, and the burr heights measured at different cutting velocity and feed are shown in Figure 7. Burr height observed is found to be substantially reduced with the use of back-up support at dry condition compared to that at experiment set I. The largest burr height seen in experiment set III is 0.64 mm. Figure 8 depicts the photographic view of burr formed using a back-up plate in dry condition. Transient burrs are seen in these tests around the hole exit end at few locations only.

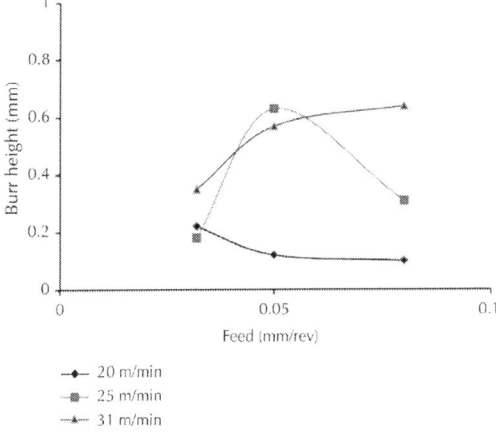

Figure 7: Plot of variation of burr height at different cutting velocity and feeds for experiment set III.

Figure 8: Photographic view of burr formed in experiment set III (left-right order: middle for Sl. numbers 1–3, top for Sl. numbers 4–6, bottom for Sl. numbers 7–9).

It is seen that burr height is quite less at low cutting velocity at all the three feeds. At low cutting velocity (V_c), when rise in machining temperature, and hence, plastic deformation is less, possibility of large burr through sustained bending is also expected to be less. Minimum burr height of only 0.10 mm is observed with the use of back-up support at dry condition at a feed of 0.08 mm/rev and cutting velocity of 20 m/min. Burr height more than 0.5 mm but less than 0.65 mm is found at some other higher cutting velocity conditions.

Discussion on Drilling Burr Formation Using Water-Cooled Condition with a Back-Up Support (Experiment Set IV)

Experiment set IV is carried out with a back-up support material under water-cooled condition. Burrs formed at these conditions are noted, and the result is shown in Figure 9. Photographic views of burrs present in this experiment are shown in Figure 10. It is found that burrs are reduced considerably under this water-cooled condition using a back-up support at 0.032 and 0.05 mm/rev feed conditions. Quite less burrs as low as 0.04 mm and 0.05 mm in height are obtained at a feed of 0.05 mm/rev and cutting velocity (V_c) of 20 and 25 m/min, respectively. However, at a higher feed of 0.08 mm/rev and cutting velocity (V_c) of 20 and 25 m/min, large burr of more than 1 mm is detected. Only at one or two points, small size burrs are found in this experiment set.

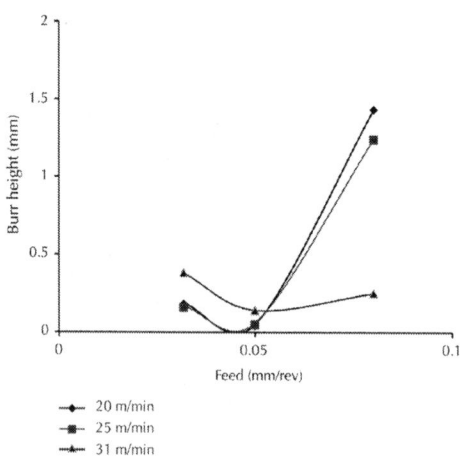

Figure 9: Plot of variation of burr height at different cutting velocity and feeds for experiment set IV.

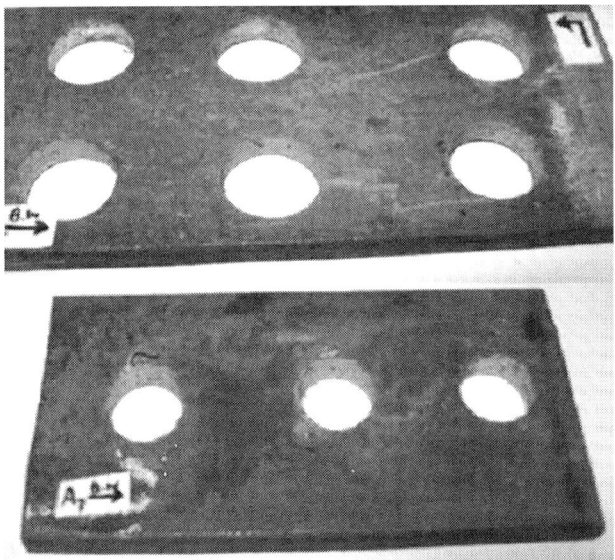

Figure 10: Photographic view of burr formed in experiment set IV (left-right order: middle for Sl. numbers 1–3, top for Sl. numbers 4–6, bottom for Sl. numbers 7–9).

Combined effect of attaching the back-up support and reduction in cutting temperature due to water cooling may have caused the formation of less burr height. The back-up support is expected to restrict the rotation of shear plane to a negative orientation as indicated in Figure 2, and to reduce burr formation significantly. Water cooling reduces cutting temperature and may have not allowed large increase in plasticity of the workpiece during machining causing less tendency of burr formation.

Discussion on Drilling Burr Formation Using Exit Edge Beveling under Dry Condition (Experiment Set V)

Variation of burr height with the change in feed at three cutting velocities is plotted in Figure 11 with the provision of 31° edge beveling in dry condition. The burr formed is observed, and its photograph is shown in Figure 12.

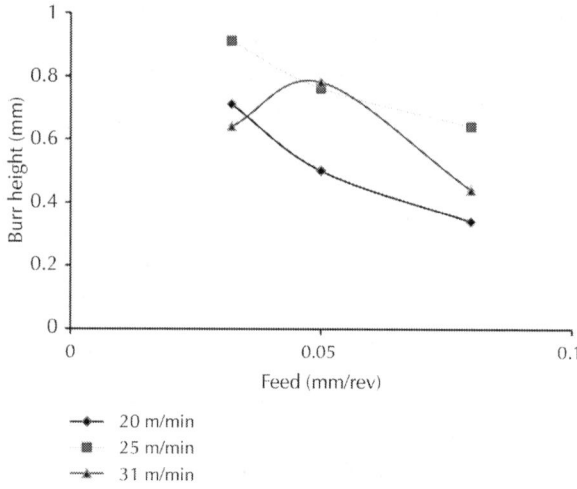

Figure 11: Plot of variation of burr height at different cutting velocity and feeds for experiment set V.

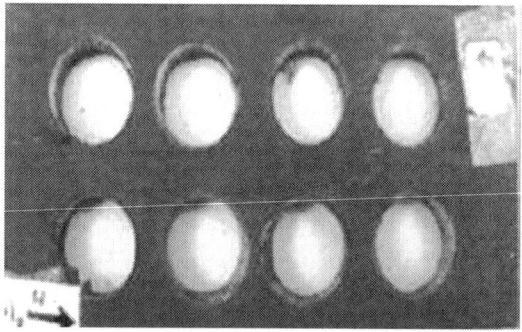

Figure 12: Photographic view of burr formed in experiment set V (left-right order: bottom for Sl. numbers 1–4, top for Sl. numbers 5–8, right side for Sl. number 9).

Moderate burr height less than 1 mm is formed at all the machining conditions undertaken in experiment set V. Burr size in this case is lower than that without using a back-up support or an edge bevel but higher than that using a back-up support. Gradual reduction of depth of cut along the bevel needs reducing force values and tends to reduce burr formation. At some points around the exit edge, few burrs are found to be attached. Further experiments are next performed using water-cooled condition to reduce temperature maintaining the exit edge bevel of 31° to investigate its effectiveness. This observation is also supported by somewhat similar reports made previously [15, 23, 27] under varied experimental conditions.

Discussion on Drilling Burr Formation Using an Exit Edge Bevel with Water Cooling (Experiment Set VI)

Experiment set VI is carried out in water-cooled condition with three values of cutting velocity and feed with the provision of an exit edge bevel of 31°. Burr formation in these conditions is also observed to be similar to the other five experiment sets. Results are presented in the form of Figure 13. Quite less burr height is observed from the plot except at 31 m/min cutting velocity and 0.032 mm/rev feed condition. The photographic view of drilling burr corresponding to experiment set VI is shown in Figure 14. Tiny infrequent burr is seen around the exit edge of the drilled hole in this experiment set.

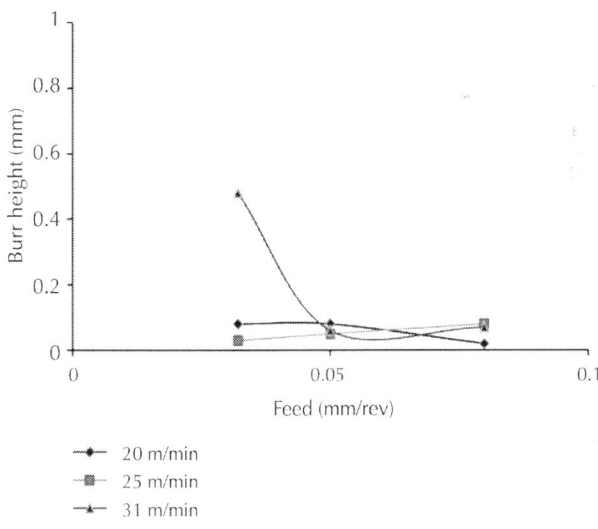

Figure 13: Plot of variation of burr height at different cutting velocity and feeds for experiment set VI.

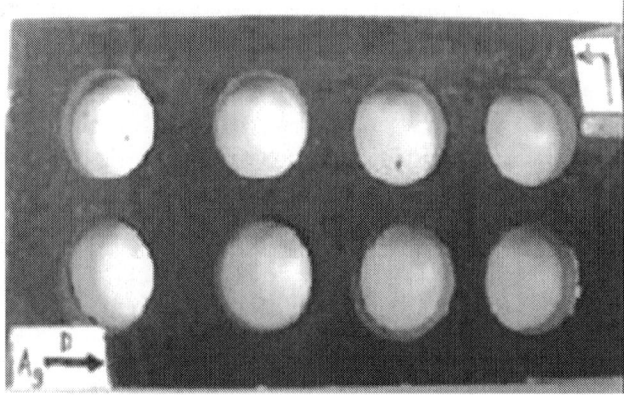

Figure 14: Photographic view of burr formed in experiment set VI (left-right order: bottom for Sl. numbers 1–4, top for Sl. numbers 5–8, right side for Sl. number 9).

Figure 14 reveals that burr is reduced substantially by using an exit edge bevel angle of 31° at water-cooled condition. Only at a cutting velocity (V_c) of 31 m/min and feed of 0.032 mm/rev, large burr height of 0.48 mm is observed. Under other experimental conditions, low burr height of up to 0.08 mm is observed. At a cutting velocity (V_c) of 20 m/min and a feed of 0.08 mm/rev, a minimum burr height of 0.02 mm is obtained. This is the smallest height of burr seen among different sets of the present experimental work, and hence, the condition corresponding to this quite low burr can be recommended to adopt.

This small burr height may have occurred mainly due to the provision of the 31° exit edge bevel that causes gradual reduction of depth of cut when the drill approaches the rear surface of the hole. This results in gradually less requirement of cutting force during tool exiting, and therefore, needing no additional back-up support during emergence of the drill from the rear side of the job. Consequently burr formation may have been suppressed significantly as negative shear plane formation becomes less likely. Water-cooled machining conditions further help in reducing plastic deformation by taking away the heat generated and reducing the extent of burr formation. This is in line with the report on the temperature effect on burr formation [23, 27].

Burr height at the exit edge is next modeled as a function of cutting conditions with a complex nonlinear algorithm, namely, artificial neural networks (ANN), or simply neural networks (NN). Detail of the NN applied in this work is given in the following sections.

THE NEURAL NETWORKS (NN) MODEL USED

There are several algorithms used in a neural network. In the present work, Levenberg-Marquardt multilayer Neural Networks (NN) with back propagation training algorithm and feed forward system [32–34] are used to model burr height using the data observed. Matlab software package with neural network toolbox is utilized in this work. In this algorithm, an iterative gradient method is employed to compute connection weights corresponding to minimum total mean-square error between the obtained output of the multilayer network and the target output. Multilayer NN consists of an input layer, one or more hidden layer(s) and an output layer [32, 33]. The neural network structure used in the present work is shown in Figure 15, where there are one input layer having five input nodes, a single hidden layer, and an output layer with one output node. Cutting velocity, feed, use of coolant, back-up support, and edge bevel comprise of the input layer, and burr height is there as the output node. The NN algorithm used in this work is detailed elsewhere [34].

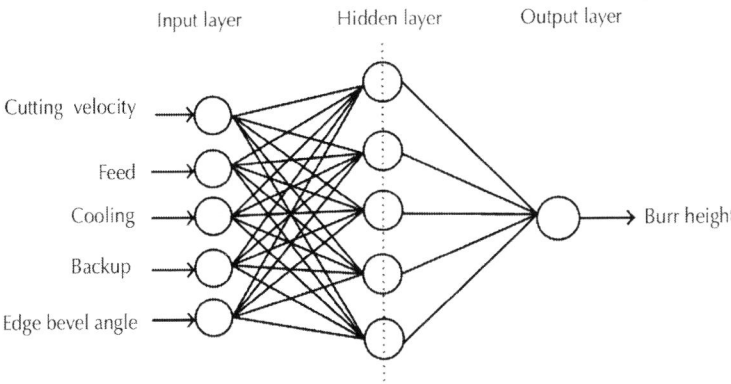

Figure 15: Architecture of artificial neural networks used.

USING THE NN FOR TESTING AND EXPERIMENTAL VALIDATION

The neural networks (NN) are first trained with a training data set. There are total 54 experimental datasets. These datasets are shown in Table 3. Train ratio chosen is 80/100 which means that 44 datasets (i.e., 80% of 54 datasets) are used for training (1st columns Sl. number 1 to 44). A validation ratio of 10/100 is used, meaning the use of 5 samples (10% of datasets) for validation (1st columns Sl. numbers 45 to 49). Testing ratio of 10/100 is selected in this work that means 5 sample data (10% of datasets) are to be used for testing (1st columns Sl. numbers 50 to 54). Training sample data are required for determining weights of the network during training. Validation of sample data is used to measure network generalization and to halt training when generalization stops improving. The testing sample data have no effect on training. They provide an independent measure of network performance during and after training. The training dataset consists of 44 sets of sample data consisting of normalized datasets of input data and the corresponding output data. The normalizing factor considered is (x_{in}/x_{max}) where x_{in} is input data and x_{max} is the maximum value data. The five input variables are cutting velocity, feed, use of coolant, use of a back-up support, and provision of exit edge bevel, and the output is experimentally observed burr height at the exit edge. For use of NN training, dry condition, use of no back-up, and no edge bevel of the work piece are assigned 0 values as the input, and water-cooled condition, use of back-up support, and 31° edge bevel angle of the workpiece are assigned a value of 1 each.

Table 3: Training dataset for neural networks (NN) and estimated burr height

Training data number	Cutting velocity, Vc (m/min)	Feed (mm/rev)	Cooling applied	Use of back up plate	Edge beveling	Measured burr height (M) (mm)	NN estimate burr height (S) (mm)	Percentage of predication error $\frac{\Delta M - \Delta S}{\Delta M} \times 100$
1	20	0.032	0	0	0	4	4.178	−4.45
2	20	0.032	1	0	0	2.73	2.794	−2.34432
3	20	0.032	0	1	0	0.22	0.214	2.727273
4	20	0.032	1	1	0	0.18	0.192	−6.66667
5	20	0.032	0	0	1	0.71	0.818	−15.2113
6	20	0.032	1	0	1	0.08	0.074	7.5
7	20	0.05	0	0	0	5.22	5.047	3.314176
8	20	0.05	1	0	0	6.17	5.938	3.76013
9	20	0.05	0	1	0	0.12	0.132	−10
10	20	0.05	1	1	0	0.04	0.0393	1.75
11	20	0.05	0	0	1	0.5	0.49	2
12	20	0.05	1	0	1	0.08	0.076	5
13	20	0.08	0	0	0	1.93	2.092	−8.39378
14	20	0.08	1	0	0	4.17	3.075	26.25899

15	20	0.08	0	1	0	0.1	0.109	-9
16	20	0.08	1	1	0	1.43	1.373	3.986014
17	20	0.08	0	0	1	0.34	0.429	-26.1765
18	20	0.08	1	0	1	0.02	0.024	-20
19	25	0.032	0	0	0	7	6.625	5.357143
20	25	0.032	1	0	0	4.76	4.894	-2.81513
21	25	0.032	0	1	0	0.18	0.219	-21.6667
22	25	0.032	1	1	0	0.16	0.159	0.625
23	25	0.032	0	0	1	0.91	0.791	13.07692
24	25	0.032	1	0	1	0.03	0.027	10
25	25	0.05	0	0	0	5.52	5.66	-2.53623
26	25	0.05	1	0	0	5.66	6.044	-6.78445
27	25	0.05	0	1	0	0.63	0.568	9.84127
28	25	0.05	1	1	0	0.05	0.041	18
29	25	0.05	0	0	1	0.76	0.664	12.63158
30	25	0.05	1	0	1	0.05	0.047	6
31	25	0.08	0	0	0	4.71	4.246	9.85138
32	25	0.08	1	0	0	3.27	3.718	-13.7003
33	25	0.08	0	1	0	0.31	0.362	-16.7742
34	25	0.08	1	1	0	1.24	1.25	-0.80645
35	25	0.08	0	0	1	0.64	0.523	18.28125
36	25	0.08	1	0	1	0.08	0.079	1.25

37	31	0.032	0	0	0	3	3.823	−27.4333
38	31	0.032	1	0	0	4.84	5.052	−4.38017
39	31	0.032	0	1	0	0.35	0.368	−5.14286
40	31	0.032	1	1	0	0.38	0.331	12.89474
41	31	0.032	0	0	1	0.64	0.732	−14.375
42	31	0.032	1	0	1	0.48	0.471	1.875
43	31	0.05	0	0	0	5.05	4.479	11.30693
44	31	0.05	1	0	0	5.88	6.255	−6.37755
45	31	0.05	0	1	0	0.57	0.472	17.19298
46	31	0.05	1	1	0	0.14	0.139	0.714286
47	31	0.05	0	0	1	0.78	0.734	5.897436
48	31	0.05	1	0	1	0.48	0.358	25.41667
49	31	0.08	0	0	0	4.2	4.64	−10.4762
50	31	0.08	1	0	0	5.31	5.27	0.753296
51	31	0.08	0	1	0	0.64	0.452	29.375
52	31	0.08	1	1	0	0.35	0.449	−28.2857
53	31	0.08	0	0	1	0.44	0.423	3.863636
54	31	0.08	1	0	1	0.07	0.079	−12.8571

ANN-based burr size modeling is done in MATLAB software package using neural network toolbox. Levenberg-Marquardt multilayer neural networks (NN) with back propagation training algorithm are employed in this work. Maximum number of epochs chosen is 1000. Initial and maximum values of mu (a factor promoting convergence of a network by a typical iterative method) are 0.001 and 10^{10} with the decreasing and increasing factors of 0.1 and 10, respectively. Minimum performance gradient (MSE) considered is 10^{-7}.

The number of hidden neurons chosen in each hidden layer is (2 × input + output).Considering 1 hidden layer, the total number of hidden neurons becomes 11 for an optimum structure of NN [33, 34]. During network analysis, it is found that in general, increasing the hidden layer from 1 to 3 results in little change; the network becomes too complex to solve the problem, and so, the number of hidden layer is usually taken as 1 and the number of hidden neurons can be chosen to be 11 as an optimum one. After training of the NN, all of the 54 sample datasets are used for getting the output. The estimated burr height, thus obtained, and the error of estimation are shown in Table 3 in the last two columns beside the input dataset.

The comparison of experimental findings and neural network estimates of burr height is displayed in Figure16, and percentages of error between experimental values and model estimates are shown in Figure 17. It is found that the NN model estimates are having quite close matching with the experimental data, barring few deviations, and showing the effectiveness of the NN algorithm for modeling the input-output system to outline the possibility of estimation of burr height within the experimental domain. Occasionally, only small deviations are observed between the estimates and measured burr height. This may be due to the inherent experimental variability of the machining system and possible existence of high degree of nonlinearity in the system. It is clear from Figures 16 and 17 that model estimates deviate more when burr height is noticeably high. In the case of experiments using back-up support, or edge bevel with the use of water cooling environment that gives remarkably less burr height, estimation error is quite less. ANN was employed in earlier works [30, 32] successfully to model and estimate burr size. In line with this, finding out the condition to reduce burr height significantly is likely to be facilitated by the proposed modeling technique.

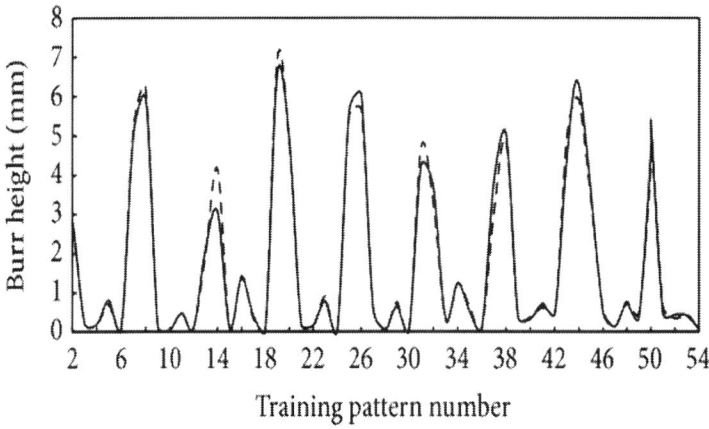

Figure 16: Experimental and ANN estimates of burr height for training patterns.

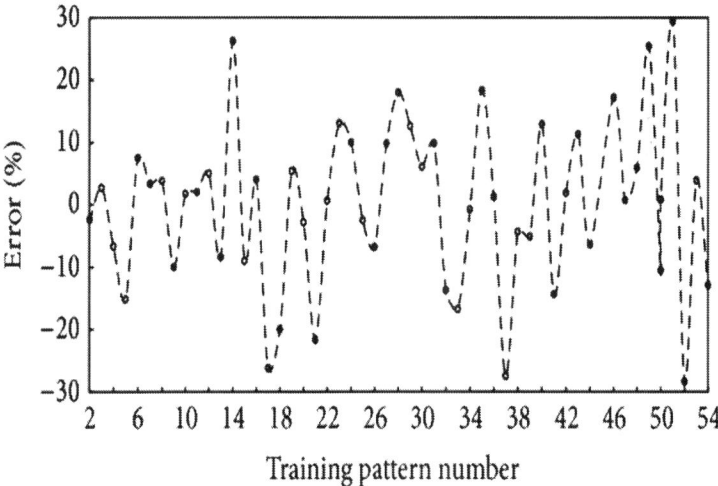

Figure 17: Experimental and ANN estimates of burr height in percentage of error.

CONCLUSIONS

From the present work with 14 mm diameter HSS twist drills for making holes in low alloy steel specimens, conclusions drawn are given below.

- Usual drilling under dry condition shows large burr formation at the exit edge. Applying water as the cutting fluid does not yield remarkable results in reducing burr height. Absence of back-up support or bevel during tool exit may have resulted in formation of large burr both in dry and wet conditions.

- Use of a back-up support reduces burr height substantially within the domain of experiments conducted due to less possibility of negative orientation of the shear plane. Use of water is found to reduce burr height further at these experimental conditions. This may be due to lowering of temperature, and hence, lessening plasticity of workpiece.

- Use of 31° exit edge bevel, made with a standard twist drill of 118° point angle, causes significant reduction in burr height in most of the cutting conditions. Use of water cooling minimizes burr height to a great extent. Slow gradual reduction of depth of cut along the bevel causes decreasing requirement of cutting force/torque needing reduced back-up support. This possibly causes suppression of burr formation to a large extent. Application of water cools down the tool and workpiece. This reduces plastic deformation resulting in further lowering of the chance of formation of large burrs.

- Within the experimental domain, hole-making at cutting velocity 20 m/min and feed 0.08 mm/rev using 14 mm drills and 31° exit edge bevel gives minimum burr under water-cooled condition. In this case, burr height as low as 0.02 mm is achieved. Hence, this condition may be recommended to obtain negligible burr.

- The three-layer neural networks algorithm is applied to model the experimental data, and the model estimates are seen to have close matching with the observed burr height with small deviations, thereby showing the possibility of using the model for estimation of burr height within the domain of experimentation.

ACKNOWLEDGMENTS

This research received no specific grant from any funding agency in the public, commercial, or not-for-profit sectors. The laboratory facilities available at Kalyani Government Engineering College are utilized for doing the work.

REFERENCES

1.   L. K. Gillespie, "Burrs produced by drilling," Tech. Rep. BDX-613-1248, Bendix Corporation, 1975.

2.   L. K. Gillespie and P. T. Blotter, "The formation and properties of machining burrs," Journal of Engineering for Industry, Transactions of the ASME, vol. 98, no. 1, pp. 66–74, 1976.

3.   K. Nakayama and M. Arai, "Burr formation in metal cutting," CIRP Annals, Manufacturing Technology, vol. 36, no. 1, pp. 33–36, 1987. ·

4.   I. W. Park and D. A. Dornfeld, "A study of burr formation mechanism," Journal of Engineering Materials and Technology, Transactions of the ASME, vol. 133, pp. 75–87, 1991.

5.   G.-L. Chern and D. A. Dornfeld, "Burr/breakout model development and experimental verification,"Journal of Engineering Materials and Technology, Transactions of the ASME, vol. 118, no. 2, pp. 201–206, 1996.

6.   S. Min, "Control chart of drilling exit burr in low carbon steel," LMA Report AISI4118, 2001.

7.   J. Kim, S. Min, and D. A. Dornfeld, "Optimization and control of drilling burr formation of AISI 304L and AISI 4118 based on drilling burr control charts," International Journal of Machine Tools and Manufacture, vol. 41, no. 7, pp. 923–936, 2001. · ·

8.   K. Lee and D. A. Dornfeld, "Micro-burr formation and minimization through process control," An e-ship Respiratory, Barkeley, Calif, USA, University of California, 2004.

9.   Y. B. Guo and D. A. Dornfeld, "Finite element analysis of drilling burr minimization with a backup material," Transactions of NAMRI/ SME, vol. 26, pp. 207–212, 1998.

10. Y. B. Guo and D. A. Dornfeld, "Finite element modeling of burr formation process in drilling 304 stainless steel," Journal of Manufacturing Science and Engineering, Transactions of the ASME, vol. 122, no. 4, pp. 612–619, 2000.

11. I. W. Park and D. A. Dornfeld, "A study of burr formation processes using the finite element method—part I," Journal of Engineering Materials and Technology, Transactions of the ASME, vol. 122, no. 2, pp. 221–228, 2000.

12. D. A. Dornfeld, "Strategies for preventing and minimizing burr formation," An e-ship Respiratory, Berkeley, Calif, USA, University of California, 2000.

13. J. C. Aurich, D. Dornfeld, P. J. Arrazola, V. Franke, L. Leitz, and S. Min, "Burrs-analysis, control and removal," CIRP Annals, Manufacturing Technology, vol. 58, no. 2, pp. 519–542, 2009. · ·

14. R. Neugebauer, G. Schmidt, and M. Dix, "Size effects in drilling burr formation," in Proceedings of the CIRP International Conference on Burrs, pp. 117–128, Kaiserslautern, Germany, 2009.

15. S. Kundu, Experimental investigation on the effect of different drilling conditions on burr formation towards its minimization [Masters Dissertation], Kalyani Government, Engineering College, Kalyani, India, 2011.

16. S.-L. Ko and J.-K. Lee, "Analysis of burr formation in drilling with a new-concept drill," Journal of Materials Processing Technology, vol. 113, no. 1–3, pp. 392–398, 2001. · ·

17. B. Shyu, "Burr reduction by tool design," LMA Research Reports, University of California, Berkeley, Calif, USA, 2001.

18. S. L. Ko and J. E. Chang, "Development of drill geometry for burr minimization in drilling," CIRP Annals, Manufacturing Technology, vol. 52, no. 1, pp. 45–48, 2003.

19. T.-R. Lin and R.-F. Shyu, "Improvement of tool life and exit burr using variable feeds when drilling stainless steel with coated drills," International Journal of Advanced Manufacturing Technology, vol. 16, no. 5, pp. 308–313, 2000. · ·

20. S. Min, D. A. Dornfeld, and Y. Nakao, "Influence of exit surface angle on drilling burr formation,"Journal of Manufacturing

Science and Engineering, Transactions of the ASME, vol. 125, no. 4, pp. 637–644, 2003. · ·

21. S. P. Pratim and S. Das, "Burr minimization in face milling: an edge bevelling approach," Proceedings of the Institution of Mechanical Engineers B: Journal of Engineering Manufacture, vol. 225, no. 9, pp. 1528–1534, 2011. · ·

22. P. P. Saha and S. Das, "An investigation on the effect of machining parameters and exit edge beveling on burr formation in milling," Journal of Mechatronics and Intelligent Manufacturing, vol. 2, no. 1/2, pp. 73–84, 2011.

23. K. Roy, P. Mukherjee, and U. K. Hansda, "An experimental investigation on drilling burr formation," inProceedings of Poster Presentation of the 3rd International and 24th AIMTDR Conference, pp. 211–216, Visakhapatnam, India, 2010.

24. S. Tripathi and D. A. Dornfeld, "Review of geometric solutions for milling burr prediction and minimization," Proceedings of the Institution of Mechanical Engineers B: Journal of Engineering Manufacture, vol. 220, no. 4, pp. 459–466, 2006.

25. K. H. Kim, C. H. Cho, S. Y. Jeon, K. Lee, and D. A. Dornfeld, "Drilling and deburring in a single process," Proceedings of the Institution of Mechanical Engineers Part B: Journal of Engineering Manufacture, vol. 217, no. 9, pp. 1327–1331, 2003. · ·

26. J. Choi, "Formation of Burr when drilling multi-layer materials," in Proceedings of the CODEF Annual Meeting, University of California, Berkeley, Calif, USA, 2003.

27. L. K. Lauderbaugh Saunders and C. A. Mauch, "An exit burr model for drilling of metals," Journal of Manufacturing Science and Engineering, Transactions of the ASME, vol. 123, no. 4, pp. 562–566, 2001.

28. J. Kim and D. A. Dornfeld, "Development of an analytical model for drilling burr formation in ductile materials," Journal of Engineering Materials and Technology, Transactions of the ASME, vol. 124, no. 2, pp. 192–198, 2002. · ·

29. U. Heisel and M. Schaal, "Burr formation in short hole drilling with minimum quantity lubrication,"Production Engineering, vol. 3, no. 2, pp. 157–163, 2009. · ·

30. S. R. Karnik, V. Gaitonde, and J. P. Davim, "Integrating Taguchi principle with genetic algorithm to minimize burr size in drilling of AISI 316L stainless steel using an artificial neural network model,"Proceedings of the Institution of Mechanical Engineers B: Journal of Engineering Manufacture, vol. 221, no. 12, pp. 1695–1704, 2007. · ·

31. Y. Nakao and Y. Watanabe, "Measurements and evaluations of drilling burr profile," Proceedings of the Institution of Mechanical Engineers B: Journal of Engineering Manufacture, vol. 220, no. 4, pp. 513–523, 2006. · ·

32. V. N. Gaitonde and S. R. Karnik, "Minimizing burr size in drilling using artificial neural network (ANN)-particle swarm optimization (PSO) approach," Journal of Intelligent Manufacturing, vol. 23, pp. 1783–1793, 2012. · ·

33. R. P. Lippmann, "An introduction to computing with neural nets," IEEE ASSP Magazine, vol. 4, no. 2, pp. 4–22, 1987. · ·

34. J.J. Moré, "The Levenberg-Marquardt algorithm: implementation and theory," Numerical Analysis: Lecture Notes in Mathematics, vol. 630, pp. 105–116, 1978.

Chapter 5

Synthesis and Performance Evaluation of a New Deoiling Agent for Treatment of Waste Oil-Based Drilling Fluids

Pingting Liu[1], Zhiyu Huang[1], Hao Deng[2], Rongsha Wang[2], and Shuixiang Xie[2]

[1]College of Chemistry and Chemical Engineering, Southwest Petroleum University, Chengdu 610500, China

[2]CNPC Research Institute of Safety & Environment Technology, Beijing 102206, China

ABSTRACT

Oil-based drilling fluid is used more and more in the field of oil and gas exploration. However, because of unrecyclable treating agent and hard treatment conditions, the traditional treating technologies of waste oil-based drilling fluid have some defects, such as waste of resource, bulky equipment, complex treatment processes, and low oil recovery rate. In this work, switchable deoiling agent (SDA), as a novel surfactant for treatment of waste oil-based drilling fluid, was synthesized by amine, formic acid, and formaldehyde solution. With this agent, the waste oil-based drilling fluid can be treated without complex process and expensive equipment. Furthermore, the agent used in the treatment can be recycled, which reduces waste of resource and energy. The switch performance, deoiling performance, structural characterization, and mechanisms of action are studied. The experimental results show that the oil content of the recycled oil is higher than 96% and more than 93% oil in waste oil-based drilling fluid can be recycled. The oil content of the solid residues of deoiling is less than 3%.

INTRODUCTION

Petroleum and natural gas are so important strategic resources that it is necessary to exploit them for all the countries. However, the large amount of waste drilling fluid, especially the waste oil-based drilling fluid which is produced in the process of drilling during field development [1, 2], is very harmful to the environment [3, 4]. Waste oil-based drilling fluid is classified as hazardous waste for the reason that it contains a lot of oil, heavy metals, and organic pollutants [5, 6]. Therefore, waste oil-based drilling fluid must be treated properly, or it will cause great harm to environment, animals and human [7, 8].

Because of the high oil content and stable emulsion of waste oil-based drilling fluid, treatment of it is different from that of the other drilling fluid and a great challenge [9]. If the oil in waste oil-based drilling fluid cannot be recycled, harmless processing will be very difficult to realize and the oil will be wasted [10]. Though some treatment technologies [11], such as thermal desorption, microwave processing, solvent extraction, chemical demulsification, and supercritical fluid extraction, have been used to treat the waste oil-based drilling fluid, it have been proved that all of them have some disadvantage [12–14]. Thermal desorption technology requires expensive equipment and high temperature, which causes high cost and energy consumption [15–17]. Microwave processing [18–20], which is an alternative to thermal desorption, also requires complex equipment and causes high cost. Solvent extraction technology and chemical demulsification technology require unrecyclable solvent or reagent to be added in the processing, which cause waste of resource [21, 22]. Though supercritical fluid is reusable, supercritical fluid extraction technology needs high temperature and high pressure [23, 24]. Therefore, it is very significant to develop more economic and effective methods to deal with waste oil-based drilling fluid.

In this paper, a novel surfactant for treatment of waste oil-based drilling fluid, called switchable deoiling agent (SDA), is synthesized to solve the problems of unrecyclable treating agent and complex processing condition. SDA is able to treat the waste oil-based drilling fluid without complex process and expensive equipment. Furthermore, SDA used in the treatment can be recycled, which overcomes the defect that the traditional agent can be used only once. Therefore, it is significant to simplify the processes and reduce waste of resource and energy by the method of using SDA.

MATERIALS AND METHODS

Materials

Sodium hydroxide, potassium hydroxide, sodium chloride, sodium sulfate, and hydrochloric acid were purchased from Beijing Chemical Works. Organic amine was purchased from Sinopharm Chemical Reagent Co., Ltd., China. Formaldehyde, formic acid, and tetrachloromethane were purchased from Xilong Chemical Co., Ltd., China. All the reagents mentioned above were of analytical reagent grade and used without further purification. Switchable deoiling agent for waste oil-based drilling fluid was synthesized. Waste oil-based drilling fluid was supplied by Daqing Oilfield in China.

Methods

Synthesis of SDA

A magnetic stirrer rotor was put into a two-neck bottle and kept rotating. Amine, formic acid, and formaldehyde solution were added in turns with the molar ratio of 1:5:2. When fog happens, two-neck bottle was cooled by ice packs to maintain relatively low temperature.

Then, condenser tube was installed on the vertical neck of the two-neck bottle and temperature probe was installed on the other neck. The temperature was maintained at 100°C for 3 h. Heating was stopped when the mixture gradually changes from light yellow to dark brown. Then, hydrochloric acid (HCl) with the same molar as amine was added when the bottle was cooled to room temperature. Heating was restarted and kept until the extra formic acid and formaldehyde were distilled out. Then, 30% sodium hydroxide (NaOH) solution was added into the two-neck bottle to adjust the pH of the liquid to 7-8 (the color of the solution would fade from dark brown to yellow). After that, the distillation equipment was installed to distill the mixture until the liquid in the bottle turns to red. After liquid distilled out was separated, a small amount of potassium hydroxide (KOH) was added to break emulsion. The separated upper layer liquid is the target product.

Pass Gas Through

The ventilation device (as shown in Figure 1) used to pump gas into mixed liquid consists of 2 parts: a 100 mL cylinder and a 10 mm diameter aeration head connected with a 4 mm diameter flexible pipe. The gas was passed through at the speed of 1.8 L/min through the pipe and aeration head. Due to its own weight, the aeration head always remains at the bottom of the cylinder, which produces uniform and fine bubbles in the liquid and makes intensive mixture of gas and liquid.

1.8 L/min

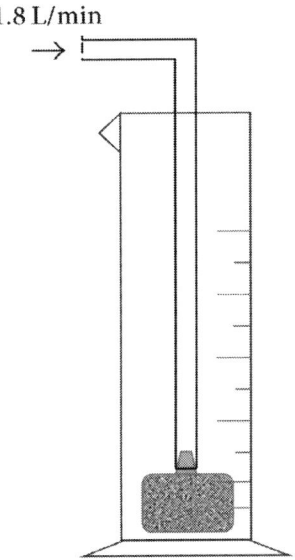

Figure 1: Diagram of ventilation device.

Determination of Oil Content in Waste Oil-Based Drilling Fluid

The water content determination apparatus, constituted by a condenser-west tube, a receiver, and the round bottomed flask, is used in this experiment. First, about 10 g (with accuracy of ±0.1 g) waste oil-based drilling fluid sample and 50 mL anhydrous petroleum ether (90–120°C) were added into the round bottomed flask. Then, the condenser-west tube and receiver were connected and the flask was heated. During the whole process, reflux rate was controlled at 2–4 drops per second. Heating was stopped when the volume of water in the receiver is no longer increased. Then, the volume of the water in the receiver was recorded. When the temperature was low enough, the rest in the flask was cleaned with anhydrous petroleum ether and filtrated with Buchner funnel. Finally, the filter residue was weighted after it was dried at the temperature of 105°C. The oil content was calculated according to the following formulas:

$$H = \frac{V_w \times \rho_w}{W} \times 100\%,$$

$$S = \frac{V_f}{W} \times 100\%,$$

$$O = 1 - H - S,$$

(1)

Where H represents the rate of water content, %; O represents the rate of oil content, %; S denotes the rate of solid content, %;denotes the weight of the sample (g); V_w is the volume of distillate (mL); ρ_w is the density of water (g/mL);W_f is the weight of filter residue (g).

Determination of Oil Content in Wastewater and Solid Waste

- *Preprocessing*: First, 10 mL acidulated water sample was mixed fully with 10 g sodium chloride (NaCl) and 20 mL tetrachloromethane (CCl_4) in a separating funnel. Then, the under-layer liquid was filtrated by a sand core funnel with 1 cm anhydrous sodium sulfate on the top. The filtrate was collected in a 50 mL volumetric flask. After that, 20 mL CCl_4 was added for the second extraction. Finally, the sand core funnel was cleaned with a small amount of CCl_4 and additional CCl_4 was added into the volumetric flask to the volume of 50 ml.

25 mL CCl_4 was added into the mixed liquid which was fully mixed with 1.00 g solid residue and 20.00 g anhydrous sodium sulfate. Then, the mixed liquid was filtrated by a sand core funnel. The filtrate was diluted to 25 mL with CCl_4 after the sand core funnel was cleaned twice with CCl_4. At last, 1 mL liquid was drawn out and diluted to 50 mL.

- *Determination by Infrared Spectrophotometry*: The oil content of liquid in 50 mL flasks was determined successively with infrared oil analyzer. The data obtained by infrared oil analyzer was multiplied by dilute multiple to obtain the oil content of oily wastewater and oily solid waste.

Structural Characterization

Fourier transform infrared spectroscopy (FTIR) (Nicolet iS50) was used for structural characterization of the deoiling agent. The samples were prepared based on pure potassium bromide (KBr) discs. First, the fully dried KBr was grinded to below 2 μm with agate mortar. After that, 70 mg grinded KBr was weighed and put into specific tableting press and then pressed to homogeneous transparent round slice under the pressure of 10 t with 5 min. Then, the pure KBr discs were impregnated into the sample solution for seconds. After that, the discs were taken out and the excessive samples were absorbed by filter paper. Finally, the prepared samples were determined with infrared spectrometer on the sample holder.

RESULTS AND DISCUSSION

Contamination in Waste Oil-Based Drilling Fluid

Five kinds of waste oil-based drilling fluid from Daqing Oilfield were used to analyze what the key components and the main pollutants were in.

As shown in Tables 1 and 2, the oil content of these waste oil-based drilling fluids is between 26.7% and 39.6%. The primary pollutant is oil. Therefore, removing and recycling the oil is the most important target for harmless treatment and resource utilization of waste oil-based drilling fluid.

Table 1: Content of each composition of waste oil-based drilling fluid

Sample source	Water content (wt%)	Solidity content (wt%)	Oil content (wt%)
Unused oil-based drilling fluid (Daqing Oilfield)	58.8	14.5	26.7

Waste oil-based drilling fluid (number 501, W.S., Daqing Oilfield)	65.8	3.3	30.9
Displacing mud (number 501, W.S., Daqing Oilfield)	47.3	13.1	39.6
Waste oil-based drilling fluid (number 11, X., Daqing Oilfield)	50.4	20.4	29.2
Displacing mud (number 11, X., Daqing Oilfield)	44.3	19.2	36.5

Table 2: Main pollutants in waste oil-based drilling fluid

Sample source	Cr (mg/ kg)	Pb (mg/ kg)	As (mg/ kg)	Hg (mg/ kg)	Cd (mg/ kg)	Oil (mg/ kg)
Unused oil-based drilling fluid (Daqing Oilfield)	18.50	14.74	14.30	1.239	0.14	267000
Waste oil-based drilling fluid (number 501, W.S., Daqing Oilfield)	10.70	21.20	20.38	0.754	0.35	309000
Displacing mud (number 501, W.S., Daqing Oilfield)	21.30	11.50	15.73	1.213	0.17	396000
Waste oil-based drilling fluid (number11, X., Daqing Oilfield)	13.60	30.70	17.92	0.836	0.41	292000
Displacing mud (number 11, X., Daqing Oilfield)	9.80	16.40	10.30	0.727	0.35	365000
Control standards for pollutants in sledges from agricultural(GB4284-84)	≤1000	≤1000	≤75	≤15	≤20	≤3000

Performance of SDA

Switch Performance of SDA

- *Switch Processes*: The main characteristic of the deoiling agent is that its hydrophilicity can be converted. Normally, the deoiling agent is hydrophobic and stratification is obvious when it is mixed with water (Figure2(a)). However, after gas A (CO_2) is bubbled into the mixtures, the deoiling agent becomes hydrophilic and the solution is homogeneous (Figure 2(b)). Again, the deoiling agent becomes hydrophobic and stratification is obvious after gas B (air, Ar, or N_2) is passed through (Figure 2(c)).

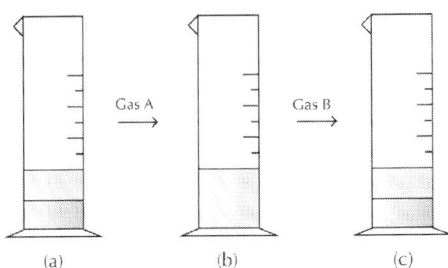

Figure 2: Switch processes.

- *Switch from Hydrophobicity to Hydrophilicity*: The experiment of switch from hydrophobicity to hydrophilicity was done under the condition of 22°C and 25% RH. First, 10 mL deoiling agent was mixed with 10 mL pure water, which formed layered liquid. Then, gas A was passed through the liquid above. Table 3shows the time of passing gas through, volume of the hydrophilic layer, volume of the hydrophobic layer, and volume of switched deoiling agent. The conversion rate was calculated based on these results. Gas was stopped when the liquid changed to be homogenous phase, which meant that the deoiling agent was switched to be hydrophilic. In the homogenous phase, the liquid-liquid interface disappeared and the mixed liquid became slightly translucent white liquid. When deionized water was added to the uniform liquid, it also remained miscible.

Table 3: Relationship between volume change of all phases and time of passing gas through when the deoiling agent switches from hydrophobicity to hydrophilicity

Time of passing gas through (min)	Volume of hydrophilic layer (mL)	Volume of hydrophobic layer (mL)	Switched volume (mL)	Conversion rate (%)
0	10	10	0	0
2.5	12.6	7.4	2.6	26
5	15	5	5	50
7.5	17.4	2.6	7.4	74
10	20	0	10	100

Figure 3 shows the relationship between conversion rate of SDA (S) and the time (t) of passing gas A through. The blue points are experiment data and the red line is the fitting curve. The function of the fitting curve is $y = x$, where y is conversion rate (S) and x is the time (t). That means that, when gas A is passed through at the speed of 1.8 L/min, the switch speed is 10%/min. Finally, the conversion rate rises to 100% after passing gas through for 10min. Therefore there is no loss in the switch process.

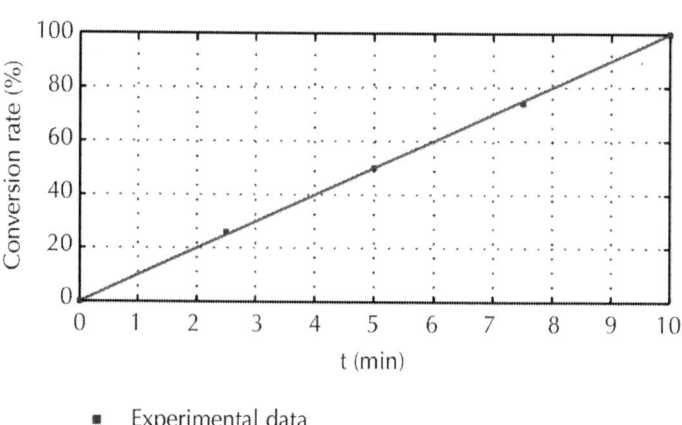

Figure 3: The fitting curve of the relationship between conversion rate of SDA and the time of passing gas A through.

- *Switch from Hydrophilicity to Hydrophobicity*: The experiment of switch from hydrophilicity to hydrophobicity was done under the condition of 22°C and 25% RH. Gas B was passed into the mixed phase (prepared in Section 2.2.1) until the volume of hydrophobic layer stopped increasing. Table 4 shows the time of passing gas B through and the corresponding volume of hydrophobic layer.

Table 4: Relationship between recovery rate of SDA and time of passing gas through when the deoiling agent switches from hydrophilicity to hydrophobicity

Gas injection time (min)	Volume of hydrophobic layer (mL)	Recovery rate (%)
0	0	0
20	1	10
40	2.5	25
60	3.9	39
80	5.1	51
100	6.4	64
120	6.5	65
140	6.5	65

The highest recovery rate is 65%. Therefore, 35% SDA is lost in the process of switching. Such high loss rate may be due to long blowing.

As shown in Table 4, the volume of oil layer increases very slightly during 100–120 min and has no change during 120–140 min, which means that almost all of the deoiling agent is switched within 120 min. Figure 4 shows the fitting curve of the recorded experiment data between 0 min and 140 min. The results show the relationship recovery rate of SDA and the time of passing gas B through. The function of the fitting curve is $y = 0.65x$ ($0 < x < 100$), $y = 65$ ($x \geq 100$), where y is recovery rate (R) and x is the time (t). This means that, when gas B is passed through at the speed of 1.8 L/min, the speed of switch is 0.65%/min.

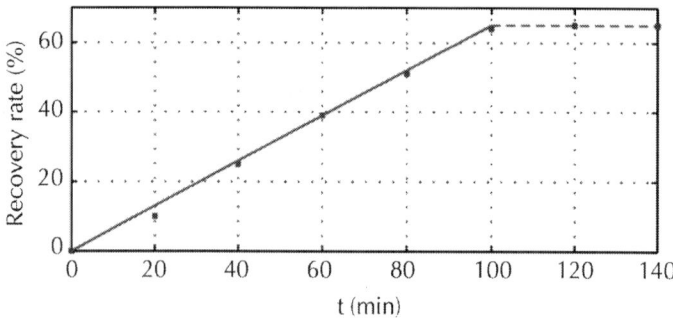

Figure 4: The fitting curve of the relationship between recovery rate of SDA and the time of passing gas A through.

In summary, the deoiling agent is hydrophobic in normal state and shows poor solubility in water. When gas A is passed through, the deoiling agent is switched to hydrophilicity and shows high water solubility. When gas B is passed through, the deoiling agent is switched back to hydrophobicity. The switching rate of deoiling agent from hydrophobicity to hydrophilicity is 15 times faster than that from hydrophobicity to hydrophilicity. Both relationships between the switched volume and the time of passing gas through in these two processes can be defined as a function y=kx, which means that the volume is proportional to the time. The experiments and analysis above show that the speed of switch from hydrophobicity to hydrophilicity is fast and the loss ratio is nearly 0%. However, the speed of switch from hydrophobicity to hydrophilicity is slow and the loss ratio is up to 35%. Therefore, there will be important researches in our future work for improving the recycle technology to reduce the loss and increasing the speed of switch from hydrophobicity to hydrophilicity.

Deoiling Performance of SDA

Figure 5 shows the process switchable deoiling agent used for removing oil from waste oil-based drilling fluid. In the first step, waste oil-based drilling fluid and SDA were mixed intensively and then stirred. The mixture was gradually divided into three layers: the upper layer was the hydrophobic layer, the mixture of SDA and oil; the intermediate layer was water; the lower layer was solid precipitation. In the second step, the solid precipitation was filtered and then gas A was passed through to switch SDA from hydrophobic to hydrophilic. In the third step, gas A was passed through continually until the layer stopped rising, which indicated that almost all of SDA was switched to hydrophilic and the oil was separated. In the fourth step, the oil was recycled and gas B was then passed through the remaining liquid. SDA was gradually switched to hydrophobic and floated upwards as oil droplet. In the fifth step, gas B was passed through until the volume of the upper layer stopped increasing. At this moment, almost all of SDA was switched back to hydrophobic and separated from the water. These SDA could be reused to deoil.

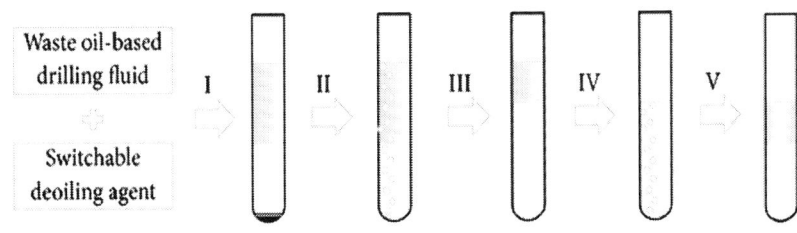

Figure 5: The process of oil removal and SDA recycle.

Sample 1 as an example was used to evaluate treatment effect. First, 41.5 g sample 1 and 100 mL switchable deoiling agent were mixed sufficiently. The mixture was sitting until clear hierarchy appeared. Solid precipitate was isolated after that. Then, 200 mL water was added into the mixture while gas A was continuously passed though. After that, the gas was shut down five min later when the bottom of the mixture was clear and droplets was flowed with gas A. At last, the separated upper black oil layer was weighed after the mixture was sat for 30 min or centrifuged for 5 min at the speed of 3000 r/min. The oil phase weighed 10.68 g. The rest of transparent mixture gradually separated into two layers after passing gas B though aqueous layer. Gas B was passed through until the volume of upper layer stopped increasing. The liquid in upper layer is recyclable switchable deoiling agent, and the lower is oily wastewater. It is found that the recycled agent can have the same nature with the new agent. The oil content of oily wastewater and oily solid waste was detected finally. The treatment effect of sample 1 was calculated as follows.

The oil content of extracted oil was calculated according to the following equations:

$$P = 1 - \frac{V_w \times \rho_w}{W} \times 100\% - \frac{W_f}{W} \times 100\%, \tag{2}$$

where P is the oil content of sample 1, %; W is the quality (g); V_w is the volume of distilled water (mL); ρ_w is the density of water (g/mL); W_f is the density of residue (g).

The percentage of oil recycle was calculated as follows:

$$R = \frac{W_{oe} \times P_{oe}}{W_{om} \times P_{om}} \times 100\%, \tag{3}$$

where R is oil recycle ratio, %; W_{oe} is the quality of extracted oil (g); P_{oe} is the content of extracted oil, %; W_{om} is the weight of waste drilling fluid (g); P_{om} is the oil content of waste drilling fluid, %.

Five kinds of waste oil-based drilling fluid were disposed of in the same way above to calculate the rate of oil removal. The related data and the result are shown in Table 5.

Table 5: The treatment result of five kinds of waste oil-based drilling fluid

Number	Weight of samples (g)	Oil content of samples (%)	Weight of extracted oil (g)	Oil content of extracted oil (%)	Oil recycle ratio (%)
1	41.5	26.7	10.68	97.3	93.8
2	35.1	30.9	10.53	96.9	94.1
3	27.4	39.6	10.40	97.8	93.7
4	36.7	29.2	10.40	97.1	94.2
5	30.1	36.5	10.64	96.8	93.7
Average	—	—	—	97.2	93.9

As the date in Table 5 showed, the switchable deoiling agent has stably good effect to dispose of waste oil-based drilling fluid. Oil contents of extracted oil are from 96.8% to 97.8%, and the average percentage is 97.2%. Oil recycle ratios are from 93.7% to 94.2% and the average percentage is 93.9%.

The oil content of oily wastewater and oily solid waste was determined by infrared spectrophotometry. The samples were extracted and diluted according to the method described in Section 2.2.3. Oil content was detected with infrared oil content analyzer (Huaxia Kechuang Oil420). Consider

$$P_w = k\phi_w,$$

$$P_s = \frac{k\phi_s}{\rho_w},$$

(4)

where P_w is oil content of oily wastewater (mg/L); K is dilution multiple; φ_w is the oil content of diluted waste water determined by infrared spectrophotometry (mg/L); Ps is oil content of oily solid waste, %; φ_s is the oil content of diluted water extracted from waste solid determined by infrared spectrophotometry (mg/L); and ρ_w is the density of tetrachloromethane (g/mL).

Five kinds of waste oil-based drilling fluid were disposed of in the same way above; the results are shown in Table 6. As shown in this table, the oil content of wastewater is between 12.83% and 16.49%, and the average percentage is 14.51%. The oil content of oily solid waste is from 1.929% to 2.915%, and the average percentage is 2.443%. It indicates that the solid waste is harmless in China. Therefore, the treatment of the solid waste would be much easier than the treatment of the original waste oil-based drilling fluid [25, 26]. These results clearly show that SDA can significantly reduce the oil content of the oil-based drilling fluid.

Table 6: The oil content of oily wastewater and oily solid waste

Number	The measured value of water (mg/L)	The measured value of solid (mg/L)	Oil content of oily wastewater (mg/L)	Oil content of oily solid waste (%)
1	73.525	33.384	14.71	2.616
2	82.437	24.617	16.49	1.929
3	66.553	37.194	13.31	2.915
4	75.968	28.752	15.19	2.253

5	64.142	31.931	12.83	2.502
Average	—	—	14.51	2.443

Structural Characterization and Mechanism of Action

Structural Characterization

Fourier transform infrared spectroscopy was used to analyze the structure of composite product (Figure 6). The location of the absorption peaks and corresponding transmittance were found after the data from spectrogram were processed.

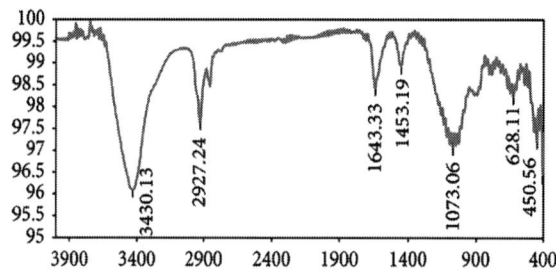

(a)

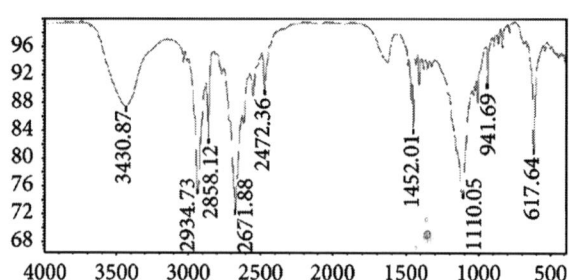

(b)

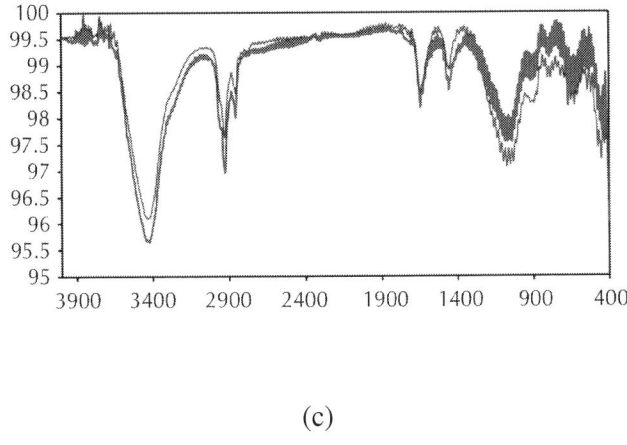

(c)

Figure 6: FTIR spectra of SDA.

Figure 6 shows the infrared spectroscopies of switchable deoiling agent in three stages of the switching progress. Figure 6(a) shows the infrared spectroscopy of original SDA, which is hydrophobic; Figure 6(b)shows the infrared spectroscopy of hydrophilic SDA switched from original SDA; Figure 6(c) shows the comparison of the infrared spectroscopy of the original SDA and the infrared spectroscopy of hydrophobic SDA switched back from hydrophilic SDA.

In Figure 6(a), the adsorption peak at $1073.06\,cm^{-1}$ has relationship with C–C bond. The two peaks at $1349\,cm^{-1}$ and $1453.19\,cm^{-1}$ come from C–H bonds in methyl. Another peak at $2927.24\,cm^{-1}$ is observed as methyl, methylene, and methane. The emergence of peak at $3430.13\,cm^{-1}$ means the existence of secondary amines or tertiary amines. According to the analysis results, the synthetic product contains large amounts of methyl, amine, and secondary carbon atoms. So it is speculated that synthetic product is based on one or several amines as the main component.

Comparison of Figures 6(a) and 6(b) shows that the position of absorption peak in Figure 6(a) is similar to the position of absorption peak is Figure 6(b). The main differences between these two figures are as follows: (1) in Figure 6(b), there is an obvious absorption peak at $2671.88\,cm^{-1}$, which means the emergence of ammonium cations; (2) in Figure 6(b), the absorption peak is pronounced weaken at $3430.87\,cm^{-1}$, which means the reduction of tertiary amine functional group. Therefore, the main change in the progress of switching hydrophobic SDA to hydrophilic SDA is that the tertiary amine is converted to ammonium salt with the effect of CO_2 (gas A) and water.

In Figure 6(c), the red curve is the infrared spectroscopy of the original SDA which is similar to the infrared spectroscopy in Figure 6(a). The blue curve is the infrared spectroscopy of the recycled SDA switched from hydrophilic SDA by passing gas B through. The blue curve shows high similarity to the red curve, which means that the recycled SDA has little difference with the original SDA and can be reused.

Mechanisms of Action

It is the main function for switchable deoiling agent, which has the characteristic of controllable hydrophilic transformation, to remove and recycle oil in the waste oil-based drilling fluid. Its main action principle is composed of deoiling mechanism and transformation mechanism.

The deoiling mechanism is based on the principle of the dissolution in the similar material structure. It refers to the similar structure and intersolubility of the solute and solvent, which means that in this paper the polar solutes dissolve in polar solvents while nonpolar solutes dissolve in nonpolar solvents. Waste oil-based drilling fluid is emulsion composed of water, oil, and solid impurities and so on. For the reason that the oil phase is mainly nonpolar and the water is polar, the oil phase is easily extracted by nonpolar solvents. And, in general, deoiling agent, which is hydrophobic, can absorb oil to realize oil-water separation. However, when gas A is passed into the mixture, oil is separated alone because deoiling agent has integrated with water for its change of hydrophilic nature. Similarly, after gas B is passed into the aqueous phase, deoiling agent is separated from water because it is hydrophobic again. So, the deoiling agent can be reused:

$$NR_3 + H_2O \underset{+\text{gas B }(-\text{gas A})}{\overset{+\text{gas A}}{\rightleftharpoons}} R_3NH^+ + (OH^- + \text{gas A})$$

$$(5)$$

The switching mechanism is that hydrophobic amine, the main composition of deoiling agent, can react reversibly according to the change of gas. As shown in the conversion mechanism (5), the hydrophobic amines (R is saturated alkyl) react with gas A to form hydrophilic product when gas A is passed into the aqueous phase. However, the reaction reverses to form original hydrophobic amine when gas B is passed into to replace gas A. Because of different reaction rate and activation energy the forward and converse reaction needs, there is great different conversion rate between the two reactions.

CONCLUSIONS

In summary, hydrophilicity of deoiling agent used for waste oil-based drilling fluid is convertible according to the need of human. Normally, deoiling agent is hydrophobic. However, it is hydrophilic when gas A (CO_2) is passed into and it is hydrophobic again when gas B (air, Ar, or N_2) is passed into. It is effective to use deoiling agent to deal with waste oil-based drilling fluid. The oil removal rate can reach 94% and the oil content of extracted oil is about 97%. The residues of deoiled waste oil-based drilling fluid are water phase and solid phase. The test result shows that the oil content of wastewater is below 17 mg/L and the oil content of oily solid waste is below 3%.

According to the analysis results of FTIR, the synthetic product contains large amounts of methyl, amine, and secondary carbon atoms. The amines especially are the leading parts. The deoiling mechanism of the switchable deoiling agent is based on the principle of the dissolution in the similar material structure. The oil is extracted from the waste oil-based drilling fluid based on the hydrophobicity of deoiling agent and the deoiling agent is recyclable for its switchable performance. The transition mechanism of it is mainly based on the reversible reaction among the amines with the water and gas. Thus, using SDA to treat the oil-based drilling waste fluid can recycle not only the oil in the fluid but also the SDA itself, which reduces waste of resource. Furthermore, with SDA, the waste oil-based drilling fluid can be treated without complex process, expensive equipment, and harsh conditions. Therefore, this technology can significantly reduce the waste of resources, energy, and the cost, comparing with the commonly used technologies.

The conversion process of deoiling agent is rapid from hydrophobic to hydrophilic and the oil loss during this process is negligible. However, the recovery process of deoiling agent, switch from hydrophilic to hydrophobic, is slow and the oil loss cannot be neglected in this process. Therefore, the next important research is how to improve the conversion rate from hydrophilic to hydrophobic and reduce the oil loss from hydrophilic to hydrophobic.

ACKNOWLEDGMENTS

The authors gratefully acknowledge the technical support of Laboratory of Solid Waste Disposal and Comprehensive Utilization of CNPC and the financial support provided by National Science and Technology Major Projects of China (2011ZX05021-004).

REFERENCES

1.   S. A. Mahmoud and M. M. Dardir, "Synthesis and evaluation of a new cationic surfactant for oil-well drilling fluid," Journal of Surfactants and Detergents, vol. 14, no. 1, pp. 123–130, 2011.
  · ·

2.   A. A. Hafiz and M. I. Abdou, "Synthesis and evaluation of polytriethanolamine monooleates for oil-based muds," Journal of Surfactants and Detergents, vol. 6, no. 3, pp. 243–251, 2003.

3.   S. Rana, "Facts and data on environmental risks-oil and gas drilling operations," in Proceedings of the SPE Asia Pacific Oil and Gas Conference and Exhibition, SPE112993, Perth, Australia, October 2008.

4.   R. Sadiq, T. Husain, B. Veitch, and N. Bose, "Risk-based decision-making for drilling waste discharges using a fuzzy synthetic evaluation technique," Ocean Engineering, vol. 31, no. 16, pp. 1929–1953, 2004. · ·

5.   T. Steliga, P. Jakubowicz, and P. Kapusta, "Changes in toxicity during in situ bioremediation of weathered drill wastes contaminated with petroleum hydrocarbons," Bioresource Technology, vol. 125, pp. 1–10, 2012. · ·

6.   I. Kisic, S. Mesic, F. Basic et al., "The effect of drilling fluids and crude oil on some chemical characteristics of soil and crops," Geoderma, vol. 149, no. 3-4, pp. 209–216, 2009. · ·

7. P. J. Cranford, D. C. Gordon Jr., K. Lee, S. L. Armsworthy, and G. H. Tremblay, "Chronic toxicity and physical disturbance effects of water- and oil-based drilling fluids and some major constituents on adult sea scallops (Placopecten magellanicus)," Marine Environmental Research, vol. 48, no. 3, pp. 225–256, 1999. · ·

8. M. Denoyelle, F. J. Jorissen, D. Martin, F. Galgani, and J. Miné, "Comparison of benthic foraminifera and macrofaunal indicators of the impact of oil-based drill mud disposal," Marine Pollution Bulletin, vol. 60, no. 11, pp. 2007–2021, 2010. · ·

9. R. S. Farinato, H. Masias, D. Garcia, R. Bingham, and G. Antle, "Separation and recycling of used oil-based drilling fluids," in Proceedings of the International Petroleum Technology Conference (IPTC ‹09), IPTC13238, Doha, Qatar, December 2009.

10. R. Sørheim, C. E. Amundsen, R. Kristiansen, and J. E. Paulsen, "Oily drill cuttings—from waste to resource," in Proceedings of the SPE International Conference on Health, Safety and Environment in Oil and Gas Exploration and Production, SPE61273, Stavanger, Norway, June 2000.

11. M. Charles and S. Sayle, "Offshore drill cuttings treatment technology evaluation," in Proceedings of the SPE International Conference on Health, Safety and Environment in Oil and Gas Exploration and Production, SPE126333, Rio de Janeiro, Brazil, April 2010.

12. J. T. Hagan, L. R. Murray, T. Meling et al., "Engineering and operational issues associated with commingled drill cuttings and produced water re-injection schemes," in Proceedings of the SPE International Conference on Health, Safety and Environment in Oil and Gas Exploration and Production, SPE73918, Kuala Lumpur, Malaysia, March 2002.

13. E. Malachosky, B. E. Shannon, J. E. Jackson, and W. G. Aubert, "Offshore disposal of oil-based drilling-fluid waste: an environmentally acceptable solution," SPE Drilling and Completion, vol. 8, no. 4, pp. 283–287, 1993.

14. H. Shang, C. E. Snape, S. W. Kingman, and J. P. Robinson, "Microwave treatment of oil-contaminated North Sea drill cuttings in a high power multimode cavity," Separation and Purification Technology, vol. 49, no. 1, pp. 84–90, 2006. · ·

15. J. T. Araruna Jr., V. L. O. Portes, A. P. L. Soares et al., "Oil spills debris clean up by thermal desorption,"Journal of Hazardous Materials, vol. 110, no. 1–3, pp. 161–171, 2004. · ·

16. R. L. Stephenson, S. Seaton, R. McCharen, E. Hernandez, and R. B. Pair, "Thermal desorption of oil from oil-based drilling fluids cuttings: processes and technologies," in Proceedings of the SPE Asia Pacific Oil and Gas Conference and Exhibition (APOGCE ‹04), SPE88486, Perth, Australia, October 2004.

17. A. J. Murray, M. Kapila, G. Ferrari, D. Degouy, B. J. Espagne, and P. Handgraaf, "Friction-based thermal desorption technology: Kashagan development project meets environmental compliance in drill-cuttings treatment and disposal," in Proceedings of the SPE Annual Technical Conference and Exhibition (ATCE ‹08), SPE116169, Denver, Colo, USA, September 2008.

18. M. S. Pereira, C. M. de Ávila Panisset, A. L. Martins, C. H. M. de Sá, M. A. de Souza Barrozoa, and C. H. Ataíde, "Microwave treatment of drilled cuttings contaminated by synthetic drilling fluid," Separation and Purification Technology, vol. 124, pp. 68–73, 2014. ·

19. J. P. Robinson, S. W. Kingman, C. E. Snape et al., "Scale-up and design of a continuous microwave treatment system for the processing of oil-contaminated drill cuttings," Chemical Engineering Research and Design, vol. 88, no. 2, pp. 146–154, 2010. · ·

20. J. P. Robinson, S. W. Kingman, C. E. Snape et al., "Remediation of oil-contaminated drill cuttings using continuous microwave heating," Chemical Engineering Journal, vol. 152, no. 2-3, pp. 458–463, 2009. · ·

21. Z. Talbi, B. Haddou, Z. Bouberka, and Z. Derriche, "Simultaneous elimination of dissolved and dispersed pollutants from cutting oil wastes using two aqueous phase extraction methods," Journal of Hazardous Materials, vol. 163, no. 2-3, pp. 748–755, 2009. · ·

22. A. A. Abdel-Azim, A. M. Abdul-Raheim, R. K. Kamel, and M. E. Abdel-Raouf, "Demulsifier systems applied to breakdown petroleum sludge," Journal of Petroleum Science and Engineering, vol. 78, no. 2, pp. 364–370, 2011. · ·

23. R. B. Eldridge, "Oil contaminant removal from drill cuttings by supercritical extraction," Industrial and Engineering Chemistry Research, vol. 35, no. 6, pp. 1901–1905, 1996.

24. C. G. Street and S. E. Guigard, "Treatment of oil-based drilling waste using supercritical carbon dioxide," Journal of Canadian Petroleum Technology, vol. 48, no. 6, pp. 26–29, 2009.

25. C.-H. Chaineau, J.-C. Setier, and A. Morillon, "Is bioremediation a solution for the treatment of oily waste?" in Proceedings of the 10th Abu Dhabi International Petroleum Exhibition and Conference, SPE78548, Abu Dhabi, UAE, 2002.

26. R. K. Dhir, L. J. Csetenyi, T. D. Dyer, and G. W. Smith, "Cleaned oil-drill cuttings for use as filler in bituminous mixtures," Construction and Building Materials, vol. 24, no. 3, pp. 322–325, 2010. · ·

Preparation and Characterization of Latex Particles as Potential Physical Shale Stabilizer in Water-Based Drilling Fluids

Junyi Liu, Zhengsong Qiu, Wei'an Huang,
Dingding Song, and Dan Bao

College of Petroleum Engineering, China University of Petroleum, Qingdao 266580, China

ABSTRACT

The poly(styrene-methyl methacrylate) latex particles as potential physical shale stabilizer were successfully synthesized with potassium persulfate as an initiator in isopropanol-water medium. The synthesized latex particles were characterized by Fourier transform infrared spectroscopy (FT-IR), particle size distribution measurement (PSD), transmission electron microscopy (TEM), and thermal gravimetric analysis (TGA). FT-IR and TGA analysis confirmed that the latex particles were prepared by polymerization of styrene and methyl methacrylate and maintained good thermal stability. TEM and PSD analysis indicated that the spherical latex particles possessed unimodal distribution from 80 nm to 345 nm with the D90 value of 276 nm. The factors influencing particle size distribution (PSD) of latex particles were also discussed in detail. The interaction between latex particles and natural shale cores was investigated quantitatively via pore pressure transmission tests. The results indicated that the latex particles as potential physical shale stabilizer could be deformable to bridge and seal the nanopores and microfractures of shale to reduce the shale permeability and prevent pore pressure transmission. What is more, the latex particles as potential physical shale stabilizer work synergistically with chemical shale stabilizer to impart superior shale stability.

INTRODUCTION

In the past decade, shale gas, as a clean and unconventional energy, has become progressively important in the energy landscape worldwide [1–5]. In drilling engineering, wellbore instability in shale, such as hole collapse, tight hole, and lost circulation, has still been a challenge especially in horizontal drilling process due to fluid penetration of water-based drilling fluids into shale matrix and subsequent pore pressure build-up and sloughing of the wellbore [6–8]. Generally, oil-based drilling fluids were chosen to drill in shale formation because of no chemical interaction between oil and shale [9]. However, the increasingly stringent environmental and economic requirements restricted its wide use and specially designed water-based drilling fluid would be an alternative. Maintaining wellbore stability using water-based drilling fluid could be achieved by sealing/consolidating wellbore physically or chemically to prevent pore pressure transmission [10, 11]. Recently, many solutions have been proposed for plugging the formation through different mechanisms, such as calcium carbonate, asphaltenes, polyglycols, and polymers, but only marginal success has been achieved [12,13]. Geologically, gas shales, a sedimentary rock, are mainly composed of clay-sized particles and are believed to be the low porosity and ultralow permeability reservoir with a significant pore volume in the nanopore range [14–16]. But conventional particles are too large to bridge and seal nanoscale pore throats and microfractures of shale and novel plugging agents in nanoscale are needed for shale stability.

Historically, emulsion polymerization was widely used in the preparation of polymer materials, especially nanomaterials, but the water resistance and surface smoothness of polymers were influenced inevitably by residual emulsifiers [17, 18]. The emulsifier-free emulsion polymerization emerged at the right moment [19,20], and polymers could be synthesized without emulsifiers or just with small amount under critical micelle concentration (CMC). But it was too difficult to control the particles size and improve the conversion rate and stability of emulsifier-free latex [21]. Recently, solvothermal method has been applied to emulsifier-free polymerization [22], and high reacting temperature and reacting pressure could make it possible to decrease particle size and improve the stability of emulsifier-free latex.

In this work, the emulsifier-free latex particles as potential physical shale stabilizer in water-based drilling fluids were synthesized by emulsifier-free emulsion polymerization of styrene (St) and methyl methacrylate (MMA) using solvothermal method with potassium persulfate as an initiator in isopropanol-water medium. The newly synthesized latex particles were characterized in detail and the influencing factors on particle size distribution (PSD) of latex particles, such as isopropanol volume, reacting temperature, and initiator concentration, were also discussed. Moreover, the interaction between latex particles and natural shale cores was investigated quantitatively via pore pressure transmission tests.

EXPERIMENTAL MATERIALS AND METHODS

Materials

The monomers of styrene (St) and methyl methacrylate (MMA), obtained from Sinopharm Chemical Reagent Co. Ltd. (China), were distilled under vacuum before use. The initiator, potassium persulfate, was obtained from Aladdin Reagent Co. Ltd. (China) and was recrystallized for purification. Isopropanol and sodium chloride, AR grade, were obtained from Sinopharm Chemical Reagent Co. Ltd. (China) and used as received. The distilled water was used throughout the experiments.

The chemical shale stabilizer, SDCS, is an aluminum complex self-developed to impart shale stability. The aluminum complex could completely dissolve in the water-based drilling fluids when pH of the fluids is maintained above 11. When interacting with formation water of low pH, it could precipitate in the shale matrix and decrease the shale permeability, thus providing a barrier to pore pressure transmission. In addition, the chemical shale stabilizer could withstand high temperature and salt concentration.

The shale samples, here, were obtained from the Sichuan Basin, China. The main clay minerals of shale samples were determined to be illite/smectite and illite by X-ray diffraction analysis (Table 1), with cation exchange capacity and surface area of 40 mmol/g and 49.65 m²/g, respectively. The cylindrical shale cores were used in the pore pressure transmission tests with a diameter of 2.54 cm and a length of 0.80 cm.

Table 1: The mineralogical composition of shale samples

X-ray diffraction	% weight
Quartz	57
K feldspar	4
Plagioclase	6
Siderite	4
Clay	29
Kaolinite	14
Chlorite	12
Illite	36
Illite/smectite	38

Preparation of P(St-MMA) Latex Particles

The P(St-MMA) latex particles were prepared by the emulsifier-free polymerization of styrene and methyl methacrylate using solvothermal method with potassium persulfate as an initiator in isopropanol-water medium. The styrene, methyl methacrylate, potassium persulfate, isopropanol, and distilled water of characteristic concentration were successively added into a hydrothermal synthesis reaction kettle. Then, the mixture above was stirred using electromagnetic stirring at room temperature for 15 min and at 90°C for another 2.5 h. The P(St-MMA) latex particles with different reaction conditions were prepared using similar methods. The recipes for synthesis of P(St-MMA) latex particles were shown in Table 2.

Table 2: Recipes of the P(St-MMA) latex particles

Number	[St] : [MMA]	[KPS]/ (mmol/L)	V (isopropanol) : V(water)	Temperature (°C)	Time (h)
S-1	1 : 1	7.40	2 : 3	70	2.5
S-2	1 : 1	7.40	2 : 3	80	2.5
S-3	1 : 1	7.40	2 : 3	90	2.5
S-4	1 : 1	7.40	2 : 3	100	2.5
S-5	1 : 1	7.40	2 : 3	90	2.0
S-6	1 : 1	7.40	2 : 3	90	3.0
S-7	1 : 1	7.40	2 : 3	90	3.5
S-8	1 : 1	7.40	2 : 3	90	4.0
S-9	1 : 1	7.40	1 : 4	90	2.5
S-10	1 : 1	7.40	3 : 7	90	2.5
S-11	1 : 1	7.40	1 : 1	90	2.5
S-13	1 : 1	5.40	2 : 3	90	2.5
S-14	1 : 1	6.40	2 : 3	90	2.5
S-15	1 : 1	10.40	2 : 3	90	2.5

Characterization of P(St-MMA) Latex Particles

The FT-IR spectra were acquired by using NEXUS 670 FT-IR spectrometer (Thermo Nicolet, USA), scanning from 4000 to 400 cm^{-1}. The purified latex particles were dried under vacuum at 80°C, and mixture of latex particles samples and potassium bromide (KBr) was pressed into pellets for FT-IR analysis. The transmission electron microscopy (TEM) images of latex particles were taken with JEOL JEM-2100UHR TEM using an accelerating voltage of 200 kV. The particle size distribution (PSD) of P(St-MMA) latex particles was analyzed with dynamic light scattering using Mastersizer 3000 (Malvern, UK). The samples of TEM and PSD were diluted before testing. The thermal gravimetric analysis (TGA) was conducted with a simultaneous thermal analyzer (NETZSCH, Germany) at a heating rate of 0~50°C/min under nitrogen atmosphere. It should be pointed out that the latex particles used for characterization were synthesized using S-3 recipe (Table 2).

Pore Pressure Transmission Tests

The pore pressure transmission test was developed using the pressure transmission technique to characterize the hydraulic properties of shale and the pore pressure transmission tests in this paper were performed on the simulation equipment for hydramechanics coupling of shale, developed by China University of Petroleum (East China) [23], and basic components of the simulation equipment are illustrated in Figure 1.

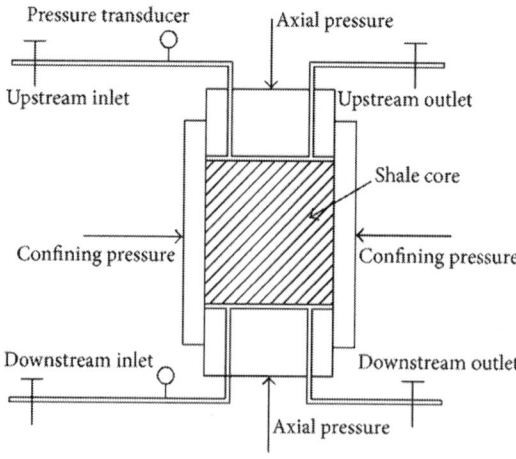

Figure 1: Schematic of pore pressure transmission test setup.

During the pore pressure transmission test, shale cores were subjected to hydraulic or osmotic gradients or both when exposed to upstream and downstream fluids. The confining pressure and axial pressure were all 5.0 MPa and upstream pressure was maintained at 3.0 MPa. The initial downstream pressure was 1.0 MPa. The pore pressure transmission tests were performed at 70°C. The downstream pressure, namely, pore pressure, was monitored throughout the tests. With no chemical potential difference between upstream and downstream fluids, the downstream pressure would become equal to the applied constant fluids pressure at the upstream because of nonzero permeability of shale cores. But, when there is chemical potential difference present, any difference between the downstream pressure and the applied constant fluid pressure could be measured in response to osmotic pressure.

The permeability of shale cores could be calculated using formula (1) [24]. Consider

$$K = \frac{\mu \beta V L}{A}$$

$$\times \left(\ln \left(\frac{P_m - P_0}{P_m - P(L, t_2)} \right) - \ln \left(\frac{P_m - P_0}{P_m - P(L, t_1)} \right) \right)$$

$$\times (t_2 - t_1)^{-1}, \tag{1}$$

where K is the permeability of shale cores, μm^2; μ is the viscosity of fluids, mPa.s; β is the static compression ratio offluids,MPa^{-1};$V$is the enclosed volume of downstreamfluids,cm^3; L is the length of shale cores, cm; A is the cross-sectionalarea, cm^2; t is total experimental time, s; P_m is the upstreampressure,MPa; P_0 is the pore pressure,MPa; $P(L, t)$ is the realtimedownstream pressure, MPa.

Shale could act as a nonideal semipermeable membrane, and membrane efficiency was defined as the ratio of actual osmotic pressure and ideal osmotic pressure to characterize its nonideality [25]. Consider

$$\sigma = \frac{\Delta P}{\Delta \Pi} = \frac{\Delta P}{(RT/V_W) \ln (a_w^{sh}/a_w^{df})}, \tag{2}$$

where σ is the membrane efficiency of shale cores, %; ΔP is the actual osmotic pressure, MPa; R is the ideal gas constant, 8.314 J·mol^{-1} ·K^{-1}; T is the test temperature, K; V W is the partial molar volume of water, 18 cm^3 ·mol^{-1}; a_w^{sh} is the water activity of pore water; a_w^{sf} is the water activity of drilling fluids.

In addition, the pore structure of shale cores before and after tests was characterized by Hitachi S-4800 field-emission scanning electron microscope (SEM) analysis.

RESULTS AND DISCUSSION

Characterization of P (St-MMA) Latex Particles

FT-IR Analysis

The FT-IR spectrum of purified latex particles (Figure 2) showed absorption peak at around 1730 cm^{-1}, corresponding to the C=O stretching band, and two characteristic absorption peaks at around 1236 cm^{-1} and 1142 cm^{-1}, corresponding to the C–O–C symmetric stretching band. The stretching vibrations of benzene skeleton were presented at around 1601 cm^{-1}, 1489 cm^{-1}, and 1454 cm^{-1}, and 754 cm^{-1} and 700 cm^{-1} were characteristic bending vibrations of single substitution benzene ring. It was also observed from Figure 2 that the C–H absorption peaks of single substitution benzene ring were found at 3093 cm^{-1}, 3065 cm^{-1}, 3020 cm^{-1}, and 2996 cm^{-1} [26, 27]. Because the P(St) and P(MMA) have been separated before testing, the above discussion confirmed that the newly synthesized latex particles were copolymers of St and MMA.

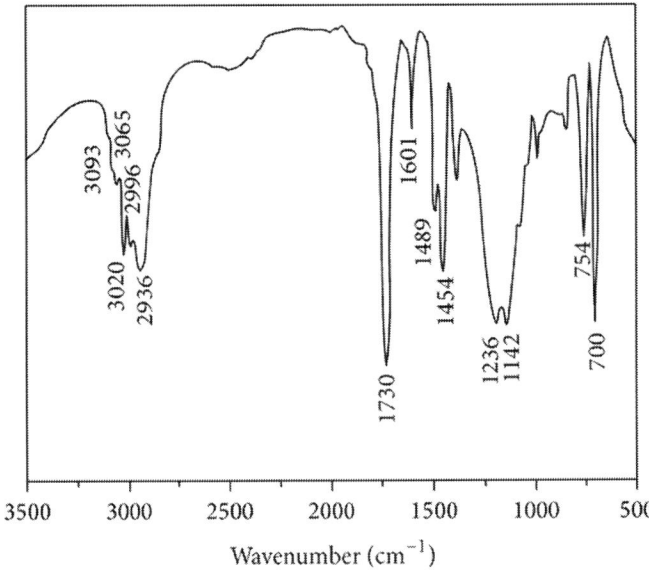

Figure 2: FT-IR spectrum of purified latex particles.

TGA Analysis

The weight loss observed up to 200°C was attributed to the desorption of physically absorbed water and dehydration of the hydrated cations. The organic compounds were decomposed between 200 and 500°C [28]. From the TGA curve (Figure 3), the purified latex particles began to decompose at around 250°C and the weight loss of latex particles would not increase significantly until temperature increased up to 380°C, indicating that the newly synthesized latex particles maintain good thermal stability. This can be attributed to the fact that the rigidity of P(St-MMA) is enhanced owing to the introduction of benzene ring [29].

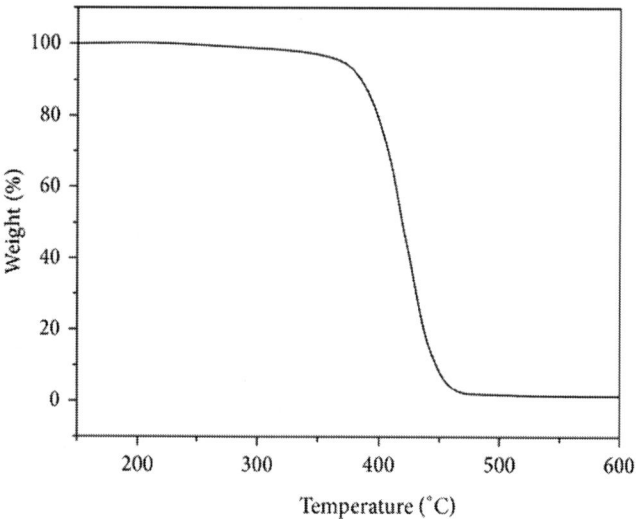

Figure 3: TGA curve of purified latex particles.

PSD and TEM Analysis

As can be seen from Figure 4, the particle size of latex particles almost unimodally distributed from 80 nm to 345 nm with the D90 value of 276 nm in addition to some aggregates with larger sizes. After entering pore throats and microfractures of shale, the coarse particles are prone to bridge or seal the largest openings of shale formation and finer particles are necessary to fill the voids between coarse particles to produce a tight immobile plug [30]. Figure 5 displays the representative TEM micrograph of P(St-MMA) latex particles. It can be seen that the latex particles are spherical and the average particle size determined by TEM is about 200 nm, which is consistent with the D50 value (208 nm) of PSD analysis.

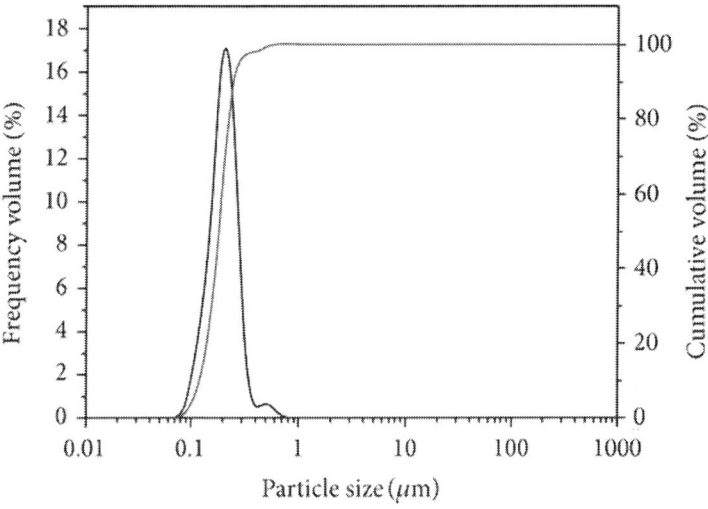

Figure 4: Particle size distribution of latex particles.

Figure 5: TEM micrograph of latex particles.

Factors Influencing PSD of P(St-MMA) Latex Particles

When it comes to reducing shale pore pressure transmission, particle size distribution is one of the most important factors influencing plugging mechanism and efficiency of pore throats and microfractures [31, 32], and D90, cumulative amount of 90% of particles which are smaller than the size, is commonly defined as a characteristic parameter. Thus, the effects of isopropanol volume, reacting temperature, and initiator concentration on particle size distribution (PSD) of latex particles were discussed here.

Isopropanol Volume

It can be seen from Figure 6 that the D90 value of latex particles decreased continuously with the increase of isopropanol volume from 10% to 50%. This can be attributed to the fact that, owing to the introduction of isopropanol, the reacting medium has lower polarity and surface tension, and it is beneficial to dispersion of droplets or particles. Additionally, isopropanol acts as chain transfer agent in the polymerization, and it could generate more surface-active oligomers in the nucleation stage and prevent particles aggregation, thus decreasing the latex particle size [33].

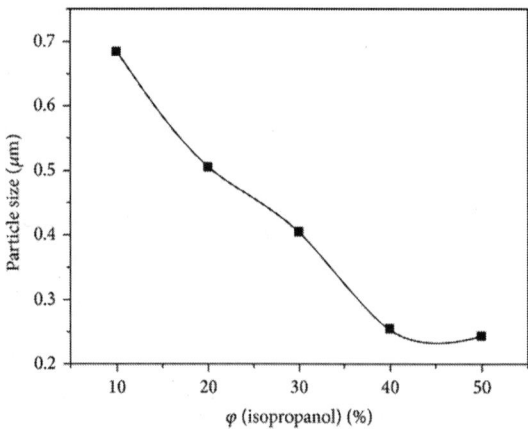

Figure 6: Effect of isopropanol volume on D90.

Reacting Temperature

The effect of reacting temperature on D90 value of latex particles is shown in Figure 7. It can be seen that the D90 value of latex particles decreased with the increase of reacting temperature, but when reacting temperature exceeded 90°C, the D90 value began to increase with the increase of reacting temperature. This can be attributed to two aspects. On the one hand, high reacting temperature increases the solubility of monomers and decomposition rate of initiators and generates more latex particles in the nucleation stage, thus decreasing the latex particle size. On the other hand, high reacting temperature could also intensify the Brownian motion and deteriorate dispersion stability of latex particles. Additionally, due to hydrophobic hydration effect and pressure effect on solubility of hydrophobic monomers, the increase of reacting temperature would also enhance the coalescence of oligomer micelle, thus increasing the latex particle size [34,35].

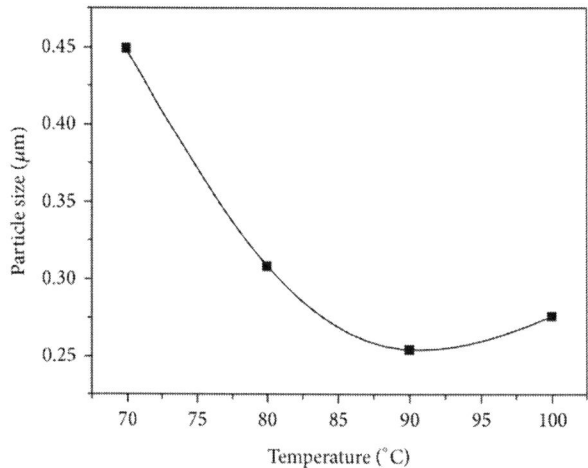

Figure 7: Effect of reacting temperature on D90.

Initiator Concentration

Figure 8 depicts the effect of initiator concentration on D90 value of latex particles. It can be seen that the D90 value of latex particles decreased with the increase of initiator concentration, and when the initiator concentration exceeded 7.4 mmol/L, the D90 value would not decrease any longer. The increase of initiator concentration makes it beneficial to generate more negatively charged surface-active oligomer micelles that are absorbed on latex particles, thus preventing latex particles aggregation and decreasing latex particles size. The high initiator concentration could also accelerate nucleation rate and polymerization rate because of high free radical production.

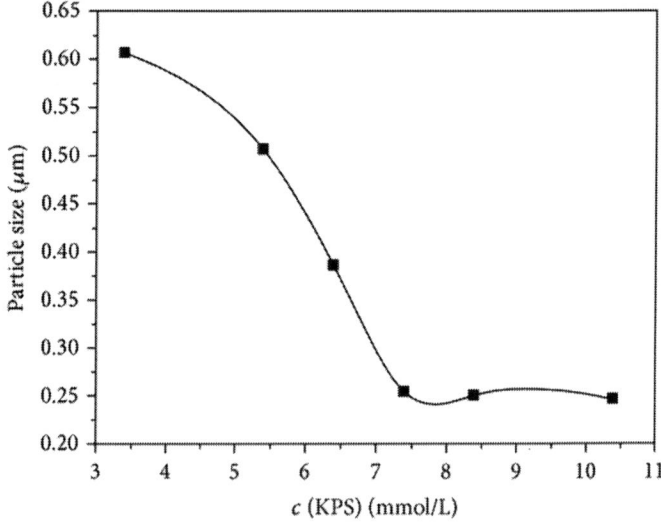

Figure 8: Effect of initiator concentration on D90.

Pore Pressure Transmission Tests

During the pore pressure transmission tests, the P (St-MMA) latex particles were used as physical shale stabilizer (SDPS) and self-developed aluminum complex was used as chemical shale stabilizer (SDCS). The downstream fluid was 3% sodium chloride solution (DSFL, $a_w = 0.998$), and the upstream fluids were 3% sodium chloride solution (USFL-1, $a_w = 0.998$), 20% sodium chloride solution (USFL-2, $a_w = 0.875$), 3% sodium chloride solution with 3% SDPS (USFL-3, $a_w = 0.998$), 20% sodium chloride solution with 2% SDPS, and 1% SDCS (USFL-4, $a_w = 0.875$), respectively.

The curves of pore pressure transmission tests are shown in Figure 9. It can be concluded that it takes more time for pore pressure build-up under hydraulic pressure when interacting with USFL-3. The permeability of shale cores could be calculated according to formula (1). After interacting with USFL-3, the permeability decreased significantly from $1.26 \times 10^{-7} \mu m^2$ (USFL-1) to $1.01 \times 10^{-8} \mu m^2$. Because there is no chemical potential difference between USFL-1/USFL-3 and DSFL, the effect of preventing or reducing pore pressure transmission is mainly related to internally bridging and sealing shale microfractures of deformable P(St-MMA) latex particles.

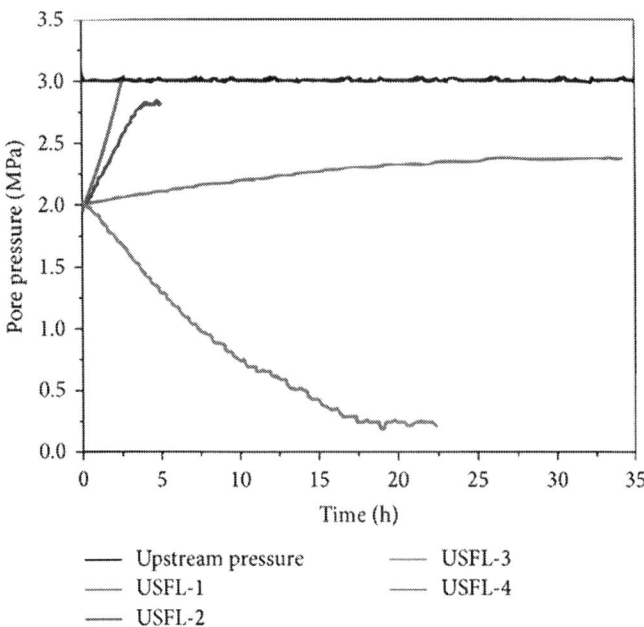

Figure 9: Pore pressure transmission test curves.

When the upstream fluid was changed to 20% sodium chloride solution (USFL-2, a_w = 0.875), the pore pressure (downstream pressure) would be less than upstream pressure due to chemical potential difference between USFL-2 and DSFL. After interacting with USDL-4, the downstream pressure decreased gradually and offset the hydraulic pressure. It is concluded that shale acts as a semipermeable membrane, and the hydraulic pressure could be offset by the backflow caused by the development of osmosis when there exists chemical potential difference. According to formula (2), the membrane efficiency of natural shale core is only 1.61%, and it increases to 14.71% after interacting with USDL-4.

As shown in Figure 10, the P(St-MMA) latex particles could be deformable to internally bridge and seal pore throats and microfractures of shale cores, and the aluminum complex would precipitate when the drilling fluid filtrate is exposed to formation water and the precipitation could block and seal pore throats and microfractures [36]. What is more, the P(St-MMA) latex particles and aluminum complex could act synergistically to reduce pore pressure transmission and increase membrane efficiency of shale cores, thus improving shale stability in drilling engineering.

S4800 5.0kV 8.0mm x300 SE(M)   100um

(a)

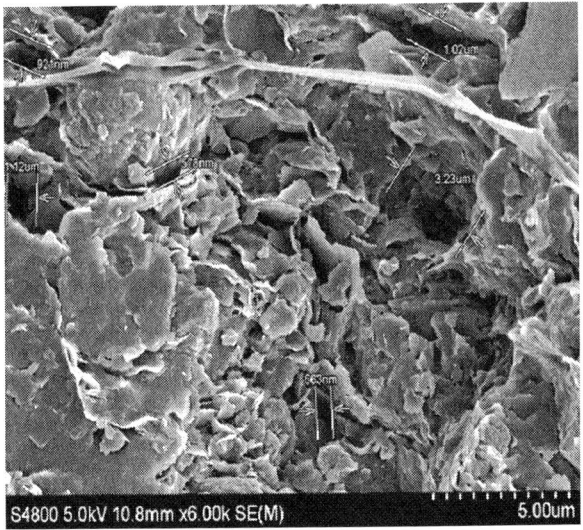

(b)

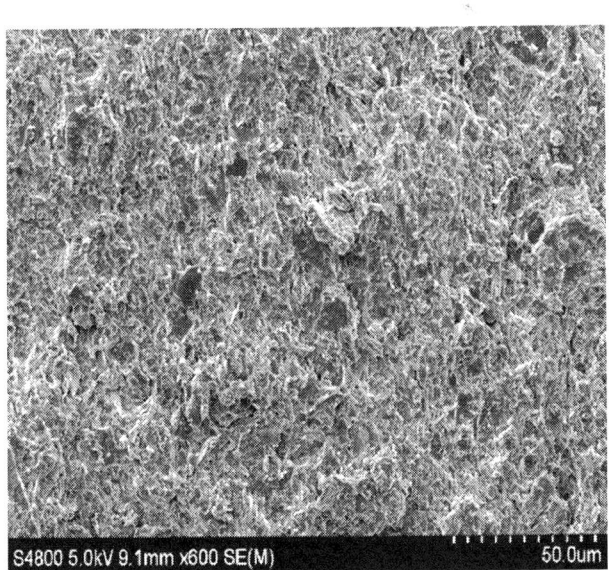

(c)

S4800 5.0kV 7.8mm x3.00k SE(M)   10.0um

(d)

Figure 10: SEM photographs of shale cores: (a)-(b) natural shale cores; (c)-(d) shale cores after interacting with USFL-4.

CONCLUSIONS

The P(St-MMA) latex particles have been successfully prepared by emulsifier-free emulsion polymerization of St and MMA, using potassium persulfate as an initiator in isopropanol-water medium. The latex particles possessed unimodal distribution from 80 nm to 345 nm with the D90 value of 276 nm, and the particle size was influenced significantly by isopropanol volume, reacting temperature, and initiator concentration. The latex particles could disperse uniformly in the water-based drilling fluids as potential physical shale stabilizer and they could be deformable to internally bridge and seal pore throats and microfractures of shale. What is more, a cooperative action was observed in P(St-MMA) latex particles (physical shale stabilizer) and aluminum complex (chemical shale stabilizer) to reduce pore pressure transmission and increase membrane efficiency of shale, thus improving shale wellbore stability. The newly synthesized latex particles were alternatives to conventional plugging agents in water-based drilling fluids for shale gas.

ACKNOWLEDGMENTS

The authors would like to acknowledge the financial support from the National Science and Technology Major Project of China (2011ZX05005-006-007HZ) and the Graduate Innovation Project of China University of Petroleum (East China) (YCX2014006).

REFERENCES

1.   K. A. Bowker, "Recent development of the Barnett Shale play, Fort Worth Basin: West Texas,"Geological Society Bulletin, vol. 42, no. 6, pp. 1–11, 2003.

2.   C. N. Zou, R. K. Zhu, B. Bai et al., "First discovery of nano-pore throat in oil and gas reservoir in China and its scientific value," Acta Petrologica Sinica, vol. 27, no. 6, pp. 1857–1864, 2011.

3.   H. B. Tan, Y. Z. Li, H. F. Tuo, M. Zhou, and B. Tian, "Experimental study on liquid/solid phase change for cold energy storage of liquefied natural gas (LNG) refrigerated vehicle," Energy, vol. 35, no. 5, pp. 1927–1935, 2010. · ·

4.   H. F. Tuo, "Thermal-economic analysis of a transcritical Rankine power cycle with reheat enhancement for a low-grade heat source," International Journal of Energy Research, vol. 37, no. 8, pp. 857–867, 2013. · ·

5.   C. R. Clarkson, J. L. Jensen, and T. A. Blasingame, "Reservoir engineering for unconventional gas reservoirs: what do we have to consider?" in Proceedings of the North American Unconventional Gas Conference and Exhibition, SPE Paper 145080, pp. 936–980, Woodlands, Tex, USA, June 2011.

6.   E. van Oort, A. H. Hale, F. K. Mody, and S. Roy, "Transport in shales and the design of improved water-based shale drilling fluids," SPE Journal, vol. 11, no. 3, pp. 137–146, 1996. · ·

7.   E. van Oort, "Physico-chemical stabilization of shales," in Proceedings of the International Symposium on Oilfield Chemistry, SPE Paper 37263, pp. 523–538, Houston, Tex, USA, February 1997.

8.   U. Tare and F. Mody, "Stabilizing boreholes while drilling reactive shales with silicate-base drilling fluids," Drilling Contract, vol. 5-6, pp. 42–44, 2000.

9.   J. C. Rojas, D. E. Clark, B. Greene, and J. G. Zhang, "Optimized salinity delivers improved drilling performance," in Proceedings of the AADE Fluids Conference, AADE-06-DF-HO-11, Houston, Tex, USA, April 2006.

10. Z. S. Qiu, J. F. Xu, K. H. Lv, et al., "A multivariate cooperation principle for wellbore stabilization," Acta Petrologica Sinica, vol. 27, no. 2, pp. 26–27, 2005.

11. J. Y. Liu, Z. S. Qiu, W. A. Huang, et al., "Experimental study on wellbore stability of fractured shale,"Eastern European Summer Time A: Energy Science and Research, vol. 32, no. 2, pp. 1213–1218, 2014.

12. M. Ramirez, D. Clapper, and P. Kenny, "Drilling-fluid design for challenging wells in the Andean Mountain region," in Proceedings of the SPE Annual Technical Conference and Exhibition, SPE Paper 102206, San Antonio, Tex, USA, September 2006.

13. S. Benaissa, D. K. Clapper, B. Hughes, P. Parigot, and D. Degouy, "Oil field applications of aluminum chemistry and experience with aluminum-based drilling fluid additive," in Proceedings of the SPE International Symposium on Oilfield Chemistry, SPE Paper 37268, Houston, Tex, USA, February 1997. · ·

14. R. J. Ambrose, R. C. Hartman, M. Diaz-Campos, I. Y. Akkutlu, and C. Sondergeld, "New pore-scale considerations for shale gas in place calculations," in Proceedings of the SPE Unconventional Gas Conference, SPE Paper 131772, Pittsburgh, Pa, USA, February 2010. · ·

15. C. R. Clarkson, N. Solano, R. M. Bustin et al., "Pore structure characterization of North American shale gas reservoirs using USANS/SANS, gas adsorption, and mercury intrusion," Fuel, vol. 103, pp. 606–616, 2013. · ·

16. C. R. Clarkson, J. M. Wood, S. D. Aquino, S. E. Burgis, M. Freeman, and V. Birss, "Nanopore structure analysis and permeability

predictions for a tight gas/shale reservoir using low-pressure adsorption and mercury intrusion techniques," in Proceedings of the SPE Americas Unconventional Resources Conference, Pittsburgh, Pa, USA, June 2012, SPE Paper 155537.

17. X.-J. Song, J. Hu, and C.-C. Wang, "Synthesis of highly surface functionalized monodispersed poly(St/DVB/GMA) nanospheres with soap-free emulsion polymerization followed by facile "click chemistry" with functionalized alkylthiols," Colloids and Surfaces A: Physicochemical and Engineering Aspects, vol. 380, no. 1–3, pp. 250–256, 2011. · ·

18. R. L. Li, C. Y. Kan, Z. P. Li, Y. Du, and Y. N. Cui, "Preparation of multihollow P(St-MAA) particles by sequence soap-free/soap emulsion polymerization and followed by stepwise alkali/acid posttreatment,"Chinese Chemical Letters, vol. 18, no. 6, pp. 741–743, 2007. · ·

19. G. Kim, S. Lim, B. H. Lee, S. E. Shim, and S. Choe, "Effect of homogeneity of methanol/water/monomer mixture on the mode of polymerization of MMA: soap-free emulsion polymerization versus dispersion polymerization," Polymer, vol. 51, no. 5, pp. 1197–1205, 2010. · ·

20. P. Bataille, M. Almassi, and M. Inoue, "Emulsifier-free emulsion polymerization of N-butyl methacrylate," Journal of Applied Polymer Science, vol. 67, no. 10, pp. 1711–1719, 1998. · ·

21. R. Yan, Y. Y. Zhang, X. H. Wang, J. Xu, D. Wang, and W. Zhang, "Synthesis of porous poly(styrene-co-acrylic acid) microspheres through one-step soap-free emulsion polymerization: whys and wherefores,"Journal of Colloid and Interface Science, vol. 368, no. 1, pp. 220–225, 2012. · ·

22. H. Q. Wu, Q. Y. Wang, D. M. Xu, et al., "Preparation of emulsifier-free cationic poly (methylmethacrylate-styrene) nanoparticles latex by solvotheraml method," Chinese Journal of Synthetic Chemistry, vol. 15, no. 6, pp. 693–696, 2007.

23. J. F. Xu and Z. S. Qiu, "Simulation test equipment of coupled hydra-mechanics of shales," Journal of China University of Petroleum (Edition of Natural Science), vol. 30, no. 3, pp. 63–66, 2006.

24. J.-F. Xu, Z.-S. Qiu, and K.-H. Lü, "Pressure transmission testing technology and simulation equipment for hydra-mechanics

coupling of shale," Acta Petrolei Sinica, vol. 26, no. 6, pp. 115–118, 2005. · ·

25. M. A. Ramirez, D. K. Clapper, and G. Sanchez, "Aluminum-based HPWBM successfully replace oil-based mud to drill exploratory wells in environmentally sensitive area," in Proceedings of the SPE Latin American and Caribbean Petroleum Engineering Conference, SPE Paper 94437, Rio de Janeiro, Brazil, June 2005.

26. W. Jiang, J. A. Joens, D. D. Dionysiou, and K. E. O'Shea, "Optimization of photocatalytic performance of TiO_2 coated glass microspheres using response surface methodology and the application for degradation of dimethyl phthalate," Journal of Photochemistry and Photobiology A: Chemistry, vol. 262, pp. 7–13, 2013. · ·

27. W. Jiang, M. Pelaez, D. D. Dionysiou, M. H. Entezari, D. Tsoutsou, and K. O'Shea, "Chromium(VI) removal by maghemite nanoparticles," Chemical Engineering Journal, vol. 222, no. 15, pp. 527–533, 2013. · ·

28. A. K. Barick and D. K. Tripathy, "Thermal and dynamic mechanical characterization of thermoplastic polyurethane/organoclay nanocomposites prepared by melt compounding," Materials Science and Engineering A, vol. 527, no. 3, pp. 812–823, 2010. · ·

29. R. Q. Bai, T. Qiu, F. Han, L. He, and X. Li, "Preparation and characterization of emulsifier-free polyphenylsilsesquioxane-poly (styrene-butyl acrylate) hybrid particles," Applied Surface Science, vol. 282, pp. 231–235, 2013. · ·

30. H. Soroush and J. H. B. Sampaio, "Investigation into strengthening methods for stabilizing wellbores in fractured formations," in Proceedings of the SPE Annual Technical Conference and Exhibition, San Antonio, Tex, USA, September 2006, SPE Paper 101802.

31. N. Kaageson-Loe, M. W. Sanders, F. Growcock, et al., "Particulate based loss-prevention material-the secrets of fracture sealing revealed," in Proceedings of the IADC/SPE Drilling Conference, SPE Paper 112595, Orlando, Fla, USA, March 2008.

32. A. Dick, T. J. Heinz, C. F. Svoboda, et al., "Optimizing the selection of bridging particles for reservoir drilling fluids," in Proceedings

of the SPE International Symposium on Formation Damage, SPE Paper 58793, Lafayette, La, USA, February 2000.

33. H. Dong, H. Q. Wu, Q. Y. Wang, et al., "Preparation of emulsifier-free poly (St-MMA) nanoparticles latex by solvotheraml method," Chinese Journal of Synthetic Chemistry, vol. 15, no. 3, pp. 330–333, 2007.

34. N. Baden, O. Kajimoto, and K. Hara, "High-pressure studies on aggregation number of surfactant micells using fluorescence quenching method," Journal of Physical Chemistry B, vol. 106, no. 34, pp. 8621–8624, 2002. ·

35. N. T. Southall, K. A. Dill, and A. D. J. Haymet, "A view of the hydrophobic effect," Journal of Physical Chemistry B, vol. 106, no. 3, pp. 521–533, 2002. · ·

36. S. F. Zhang, Z. S. Qiu, W. A. Huang, J. Cao, and X. Luo, "Characterization of a novel aluminum-based shale stabilizer," Journal of Petroleum Science and Engineering, vol. 103, pp. 36–40, 2013. · ·

Chapter 7

Assessing Level and Effectiveness of Corrosion Education in the UAE

Hwee Ling Lim

Arts & Sciences Program, The Petroleum Institute, Sas Al Nakheel Campus, Abu Dhabi, United Arab Emirates

ABSTRACT

The consequences of corrosion can be minimized by an engineering workforce well trained in corrosion fundamentals and management. Since the United Arab Emirates incurs the second highest cost of corrosion after Saudi Arabia, this paper examined the quality of corrosion education in the UAE. Surveys with academia and industry respondents showed that dedicated corrosion courses and engineering courses that integrated corrosion into the curricula were available in UAE universities, but graduates had insufficient knowledge of corrosion engineering and superficial understanding of corrosion in real-life design contexts. The effectiveness of corrosion education is determined by both competence in corrosion knowledge/skills and availability of resources (faculty and research). Though most departments would not hire new corrosion-specialist faculty, department research efforts and industry partnerships in corrosion research were present. The paper concluded with recommendations for improving knowledge and skills of future engineers in corrosion and enhancing corrosion instruction to better meet industry needs.

INTRODUCTION

The annual cost of corrosion prevention and damage worldwide is estimated at US\$ 2.2 trillion (2010). In the United Arab Emirates (UAE), the annual cost of corrosion is US\$ 14.26 billion (2011 estimate) with most of the cost related to corrosion in its energy industry. Hence, it is vital that engineers in the oil and gas industry are technically competent in the control and management of corrosion. Much of this competence is gained through formal higher education. However, there is little research on the quality of corrosion education in the UAE.

This study aims to assess the quality of corrosion education in engineering programs of universities in the UAE. The research questions posed in this study are stated below.

- RQ 1a: What is the level of corrosion instruction available in universities in the UAE?

- RQ 1b: Why is such a level of corrosion instruction available in UAE universities?

- RQ 2a: How competent are the engineering undergraduates/ graduates from UAE universities in corrosion knowledge and skills?

- RQ 2b: What resources are available to UAE universities to support effective corrosion education?

Findings from this paper could help higher education institutions in the UAE identify possible shortcomings in their corrosion education programs. Accordingly, this paper has the potential to improve the knowledge base and skills of future engineers in the recognition and management of corrosion. Also, companies in the industrial sectors where corrosion has great impact would gain much needed information on the current quality of corrosion education in the engineering programs of UAE universities. Moreover, the recommendations provided could guide higher education institutions in enhancing their existing corrosion education programs to meet industry expectations better.

LITERATURE REVIEW

Corrosion: Definition and Causes

Corrosion is defined as the deterioration of metals or materials by chemical, biological or environmental agents [1, 2]. It is a natural process involving the electrochemical oxidation of metals by an oxidant such as oxygen. "Rusting" is a typical corrosion process that is the formation of an oxide of iron due to oxidation of iron [3]. In the petrochemical refining industry, corrosion is caused by various natural constituents present in oil such as sulphur compounds, salt water, inorganic and organic acids, and nitrogen that form cyanides. In most oil and gas environments, CO_2 and H_2S are mainly responsible for corrosion [4]. CO_2 dissolves in water and forms carbonic acid that lowers the pH of the water, thus increasing its corrosiveness.

The oil/gas industry typically operates in very harsh environments such as deserts and deep seas. In such environments, corrosion is a very serious and challenging problem. Desert conditions can promote corrosion due to the presence of the following: alternating wet-dry weather cycles, high temperatures (around 50°C) and high humidity for most of the year, erosion from strong winds and blowing sand; salinity of ground water, and Sabkha soil that causes discoloration of stainless steel (316) in three years, while in nondesert conditions there is typically no discoloration in 25 years.

Also, equipment in deep sea environments, such as off-shore oil rigs and platforms, is constantly exposed to a wide range of environmental conditions; the equipment is particularly vulnerable to corrosion. In off-shore environment, corrosion severity can be related to location of equipment with respect to the ocean surface. For an oil platform, there are the atmospheric zone (degree of rainfall, sun and wind), splash zone (area wetted by wave action), and immersed zone (area below the water surface) [5]. The constant mixture of saltwater contact and oxygen interaction causes drilling tools and oil rig metallic supports to be more liable to corrosion [6]. In particular, H_2S in seawater is dangerously effective at corroding metals that can lead to the sinking of supports of the rigs into the seabed. Corrosion not only causes deterioration of materials but also poses economic and engineering problems for industries and countries.

Impact of Corrosion: Financial Cost to the GCC States

The annual cost of corrosion worldwide is estimated at US$ 2.2 trillion (2010), which is about 3% of the world's gross domestic product (GDP) of US$ 73.33 trillion [7]. The UAE annual cost of corrosion is estimated at US$ 14.26 billion (2011), which is about 5.2% of the country's GDP over three years (2009–2011). Among the GCC states, the UAE incurs the second highest annual cost of corrosion after Saudi Arabia (Figure 1).

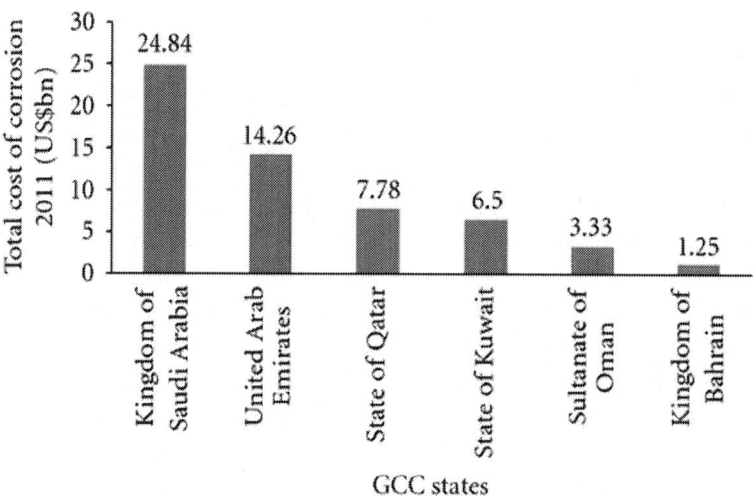

Figure 1: Annual cost of corrosion for GCC states (US$ bn), 2011 estimate [7].

The economies of the GCC states are highly dependent on the oil/gas industrial sectors. In particular, the UAE is estimated to account for

6.17% of Middle-East regional oil demand by 2014, while providing 10.87% of supply. It also contributed 10.5% to 2010 regional gas production and is estimated to provide 10.19% of supply by 2014 [8]. Most of the costs related to corrosion are incurred in the following areas:

- degradation of the pipeline infrastructure for transportation and storage of natural gas, crude oil, and refined petroleum products,
- corrosion of equipment involved in the production of corrosive crude oil from aging reservoirs,
- lost production due to downtime for corrosion inspection,
- requirement for extensive chemical treatment to maintain the integrity of assets.

Corrosion-related failures constitute 25% of failures experienced in the oil/gas industry [4]. In five Saudi Aramco refineries, corrosion accounted for 13% of engineering costs and 36% of maintenance costs from 1999 to 2001 [9]. Moreover, the cost of shutting down units and plants can reach US$ 20,000 to US$ 50,000 per hour [3].

HSE Impact of Corrosion

The impact of corrosion is not only monetary but involves environmental, health, and safety risks as well. When poorly managed, corrosion affects equipment integrity and serviceability raising the risks of leaks and discharge of flammable fluids and gases that present serious health, safety, and environmental (HSE) hazards for which companies can be held liable [10]. For example, pipelines leaks can contaminate the nearby water sources and air. Also corrosion of plant equipment housing flammable materials can lead to fire hazards. In Rawdhatain, Kuwait (January 31, 2002), a pipe leak resulted in major explosions at an oil gathering center killing four people and three main facilities were destroyed; in Al-Ahmadi, Kuwait (June 25, 2000), a condensate line between an NGL plant and refinery failed in Kuwait's largest oil refinery. Five were killed and three crude units were damaged. The cause was an aging pipe that suffered erosion-corrosion and was not detected during inspection. Given the economic and HSE impact of corrosion on the oil/gas industry, professional knowledge in prevention and management of this problem is crucial. Hence, it is important to stress the role of corrosion education for engineers.

Importance of Corrosion Education for Engineers

As it is impossible to eliminate corrosion, corrosion management is the most feasible solution. The expertise of a corrosion engineer is required for managing corrosion, especially in high-temperature and high-pressure environments [6]. Given the importance of corrosion, it is difficult to understand why 54% of corrosion protection practitioners have not taken a course on corrosion during their formal education. Also, 45% of the currently active and experienced corrosion technologists are likely to retire in the next 10 years [10]. Hence, implementing a robust corrosion education system would help the oil/gas industry to reduce the long-term cost of corrosion; decrease the potential hazards from corrosion, and develop new corrosion-resistant materials, technologies, and processes for corrosion management.

Assessing Quality of Corrosion Education

Recently, in response to the threat corrosion poses to the national security infrastructure, the Department of Defense (DoD) commissioned a National Research Council (NRC) report to assess the state of corrosion engineering education in the United States. A Committee on Assessing Corrosion Education was formed to study the situation and reported its findings in 2009 [11]. The committee was commissioned to do the following:

- assess the level and effectiveness of existing engineering curricula in corrosion science and technology, including corrosion prevention and control,
- recommend actions that could enhance the corrosion-based skill and knowledge base of graduating and practicing engineers [11, page 2].

The committee invited 83 US engineering schools to participate in an online survey and 31 institutions responded (37% response rate). Out of the 31 returns, 61% were respondents from material science departments while the rest were from other departments. The report acknowledged that since the respondents were self-selected and largely institutions already involved in corrosion education, the results probably presented an overly positive picture of the real situation and should be interpreted with some caution.

The findings showed that corrosion is taught mainly as a part of other courses in responding universities rather than as a specific course. When a dedicated corrosion course is available, it is likely to be an elective offered to undergraduates and graduates. However, when corrosion is taught (integrated) in other courses, the courses are mostly required and offered at undergraduate level with only a few lectures spent on the discussion of corrosion. Moreover, dedicated corrosion courses are likely to have lower enrolment numbers (smaller class size) compared to courses that integrate corrosion into their syllabi. Since the committee presented the most recent large scale survey on the current state of corrosion education in higher education, this study replicates in part its questionnaire design for a more meaningful comparison of findings.

Summary of the Review

Although corrosion is an inevitable natural process, the economic, environmental, and societal consequences of corrosion failure in oil/gas operations can be minimized and managed by an engineering workforce that is properly educated in corrosion fundamentals and adequately trained in corrosion management. Moreover, a knowledgeable workforce can contribute to developments in corrosion prevention technologies that are vital to improving efficiency in energy production. Findings from the National Research Council report [11] provided an up-to-date description of corrosion education available in the USA but there is currently no equivalent study available in the UAE. With the cost of corrosion increasing steadily, there is an urgent need for research on the quality of corrosion engineering education in the UAE. The next section describes the methodological design of this study.

THE CASE RESEARCH METHODOLOGY

The Single-Embedded Case: Units of Analysis and Sampling Strategy

This research used a single-embedded case design whereby within a single case, attention also is given to subunits [12]. The single-embedded case design presents the advantage of in-depth analyses of subunits that provide insight into the main case. Since this qualitative study aims to examine the quality of corrosion education in engineering programs of UAE universities, the main case is the engineering education program while four universities are subunits nested within the primary case (Figure 2).

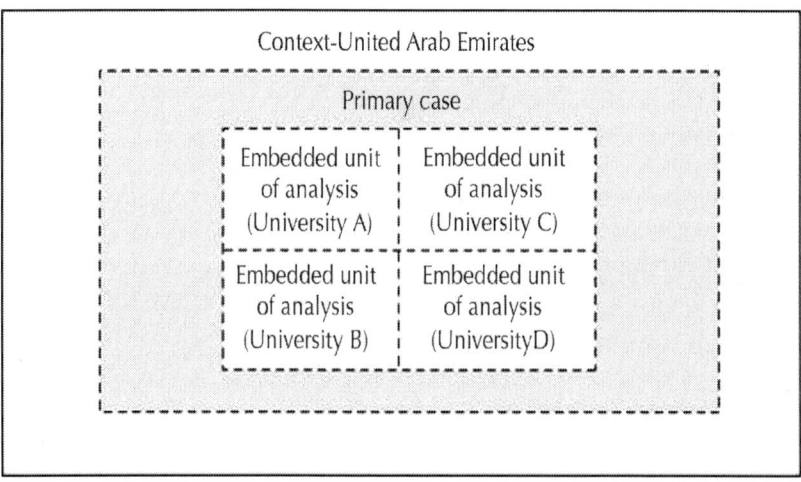

Figure 2: Single-embedded case design in this study (adapted from [13, page 40]).

A purposive sampling approach was adopted and the following criteria were used in subunit selection:

- location of the higher education institution in the UAE,
- availability of engineering degree programs in the institution,

- availability of engineering programs in which corrosion is relevant (this excludes institutions that offer engineering programs such as software/information-technology only),
- the higher education institution must have graduated at least one cohort of students,
- reliable accessibility and cooperation from universities/ participants for research.

Four institutions were identified that met all the criteria. Three institutions are located in the emirate of Abu Dhabi and one institution is in Sharjah. Due to the ethical consideration of confidentiality, the universities were pseudonymised (as University A, B, C, D), and identifying information was removed in the reporting of this study.

Data Sources

The primary data sources were academics from the four educational institutions and respondents from the oil/gas industry. The total sample size was 66 respondents. Respondents from academia (n=58) included faculty, teaching assistants, and laboratory instructors. Also, engineers (n=8) from a major oil/gas company were interviewed to understand employers' perspective on the competence of engineering graduates in corrosion. The secondary data sources were the organizations, namely, university prospectuses and websites that provided archival data on the engineering programs and curricula available in the institutions.

Instruments

A survey questionnaire (Questionnaire A) and an interview questionnaire (Questionnaire B-1 and B-2) were used. Questionnaires A and B-1 were self-administered surveys (i.e., participants completed the questionnaires on their own) that were delivered to respondents via email or face-to-face. However, Questionnaire B-2 was administered via face-to-face interviews (individual or focus group) depending on the participants' preferences. The focus group size ranged from 2 to 4 respondents per group. Interviews were used to follow up and elicit elaborations on responses provided in Questionnaire B-1 that were completed earlier by participants. Also, all interviews were audio recorded to ensure that the views of the respondents were accurately represented. Table 1 summarizes the instruments and respondent distribution.

Table 1: Instruments and respondent distribution

Subunit	Survey		Interview			
	Face-to face	E-mail	Individual (Academia)	Focus Group (Academia)	Focus Group (Industry)	Total
University A	n=10	—	—	—	—	10
University B	—	n=9	—	1 FG* (n=3)	—	12
University C	n=11	—	n=2	1 FG (n=3)	—	16
University D	n=9	—		3 FG x 3 Resp 1 FG x 2 Resp (n=11)	—	20
—	—	—	—	—	2 FG x 4 Resp* (n=8)	8
Total	39	2	17	8		66

Constructs and Measures

The questionnaires had five main parts (Parts A-E) and 37 questions (closed and open-ended) that captured information on respondent demographics, level of existing engineering curricula in corrosion, effectiveness of engineering curricula in corrosion, and recommendations for enhancing corrosion knowledge and skills. Table2 summarizes the constructs and questions that constitute the operationalized measures of the constructs.

Table 2: Measures for constructs

Construct	Measures as survey questions*
Level of existing engineering curricula in corrosion	Q10. Does your department offer a course or courses specifically in corrosion?
	(a) Yes, my department offers course(s) specifically in corrosion (b) No, my department teaches corrosion as part of other courses
	(c) No, corrosion is NOT taught at all in my department
Effectiveness of existing engineering curricula in corrosion	
Level of competence in corrosion knowledge and skills	Q34. Do you think engineering undergraduates/ graduates (from your university) have sufficient fundamental knowledge of corrosion engineering? Please explain.*
	Q35. Do you think engineering undergraduates/ graduates (from your university) have sufficient understanding of the importance of corrosion in engineering design? Please explain.*
Availability of resources (manpower, knowledge)	Q29. Would your department consider hiring a faculty member whose technical focus is corrosion?
	(a) Yes
	(b) No Q32. Is your department doing any corrosion-specific research? If so, who is funding it?
	(a) No
	(b) Yes, funding organization(s): ——————— Q33. Do you have any actual or potential partnerships with industry to study corrosion or develop continuing education for practicing engineers?
	(a) No
	(b) Yes
	f so, please describe your partnerships with industry to study corrosion

*Industry respondents answered a variation of Q34/35 that excluded reference to "from your university."

The construct level of existing engineering curricula in corrosion is defined as the extent of corrosion instruction available. The construct is operationalized by a question (Q.10) that measures availability of corrosion instruction on a scale from having a dedicated corrosion course, teaching corrosion as a part of other courses (integrated), to not teaching corrosion at all.

The construct effectiveness of existing engineering curricula in corrosion is defined as the degree to which the course produces the desired instructional outcomes measured as

- level of competence in corrosion knowledge and skills of engineering undergraduates/graduates (Q.34-35),
- availability of resources to achieve the desired outcomes (Q.29, 32-33).

Data Analysis and Validity Issues

Quantitative and qualitative data were gathered, respectively, from closed and open-ended questions in the instruments. While the quantitative data was presented as descriptive statistics, qualitative responses in the survey and interview datasets were subjected to interpretive content analysis to derive main themes.

The reliability of a study entails the demonstration that operations can be repeated with the same results. This study acknowledges subjectivity with the use of a single case and that knowledge gained is based on interpretations in a specific time/context and not be claimed to be generalizable to whole populations. However, the standard of authenticity [14] is adopted and the qualitative study's methodological design emphasized the use of multiple data sources and methods (Sections 3.2 and 3.3) for capturing "rich" descriptions that add to credibility of the findings. Diverse perspectives from academia and industry participants were gathered for triangulation rather than the confirmation of a single interpretation of the quality of corrosion education in the universities.

While external validity, as generalizability of the findings, is moot in this single case study, internal validity is relevant in terms of content and construct validity of the instruments. Both issues were addressed by piloting and refining the instruments during which questions were added, discarded, rephrased, and reordered that improved clarity and layout of the instruments.

Ethical Considerations

Research ethics is the application of fundamental and moral principles of respect for autonomy, beneficence, and justice in doing research [15]. In this study, ethical issues of informed consent, voluntary participation, and confidentiality were relevant. Informed consent refers to the necessity of providing respondents with an understanding participation of the vouluntary, the purpose of the research, anticipated consequences of participation and factors which could reasonably influence their decision. Voluntary participation refers to the right to privacy and the right not to answer specific questions. Since forced participation may cause respondents to withhold vital information, obtaining voluntary participation is critical. Confidentiality means that the real identities of participants would not be revealed in the dissemination of findings. Being professionals in their respective fields, any direct linkage of participants' personal views to their actual identities may jeopardize their positions in the institutions. These issues were handled by a cover letter and informed consent form that were read, understood, and signed by participants prior to taking part. Confidentiality was ensured by deleting compromising details and replacing actual names and identities of employer institutions with pseudonyms in data processing and publication.

RESULTS AND DISCUSSION

Overview of Institutions and Respondents

Most of the academia respondents were male, aged 31–50 years (67%), with more than 10 years work experience. Most had higher education degrees (Ph.D., M.S. degrees), held positions from professors to lab instructors in various departments (mechanical, civil, and chemical engineering). The industry respondents were all employed by a major oil/gas company, most were also male but younger (75% below 40 years), with more than 10 years work experience. Most had bachelor's degrees and in positions ranging from heads of engineering divisions to senior/specialist engineers.

RQ 1a: What Is the Level of Corrosion Instruction Available in Universities in the UAE?

The level of corrosion instruction available is measured on a scale from having a dedicated corrosion course, teaching corrosion as a part of other courses (integrated), and not teaching corrosion at all. Although the departments in University A that were surveyed offered engineering programs in which corrosion would be relevant (civil engineering, master of engineering management), it was found that corrosion was not taught at all by the departments.

The remaining three universities offered dedicated corrosion courses and taught corrosion as a part of other courses as well (Table 3). The dedicated corrosion courses were mainly electives, offered once a year at undergraduate level, included laboratory work and class size ranged from 20 to 50 students (Table 4). Thecourses that integrated corrosion into their curricula were offered at undergraduate and graduate levels (mixed) and only a few lectures were devoted to corrosion (Table 5).

Table 3: Level of existing engineering corrosion curricula available

Q. 10	University A	University B	University C	University D
Dedicated course(s) in corrosion	—	Yes	Yes	Yes
Corrosion covered in other courses	—	Yes	Yes	Yes
Corrosion not taught at all	Yes	—	—	—

Table 4: Details on dedicated corrosion courses

Dedicated corrosion courses				
	University A	University B	University C	University D
Class level	—	Undergraduate	Undergraduate	Mixed*
Class size*	—	Fewer than 20 20–50	20–50	Fewer than 20 20–50
Required or elective	—	Required	Elective	Elective
Availability**	—	Once a year	Once a year	Every semester Once a year
Lab-based course	—	Available	Available	Available
Course name#	—	(i) Corrosion	(i) Corrosion	(i) Material science
		(ii) Introduction to corrosion		(ii) Corrosion engineering
		(iii) Materials and corrosion		(iii) Advanced physical chemistry
				(iv) Corrosion chemistry

*Class size: number of students enrolled per class. **Availability: some of the courses were available once a year, others were available every semester. #Course name: courses listed based on participants' own interpretations of how much corrosion content was taught in the course.

Table 5: Details on courses that integrate corrosion in curricula

Courses that integrate corrosion into curricula				
	University A	University B	University C	University D
Class level	—	Mixed*	Mixed	Mixed
Time devoted to corrosion	—	A few lectures	A few lectures	A few lectures

Course name		(i) Civil engineering construction	(i) Material sciences	(i) Production engineering
	—	(ii) Construction materials	(ii) Construction materials	(ii) Production and surface facilities
		(iii) Materials and corrosion	(iii) Physical chemistry	(iii) Mechanics of materials
		(iv) Chemistry for civil engineers		(iv) Failure analysis and prevention

*Mixed: undergraduate and graduate levels.

Compared to the NRC report [11], the structure of dedicated and integrated corrosion courses were similar except in two minor areas. In the UAE universities, the dedicated corrosion course class size tended to be larger (20–50 students per class) and the integrated courses were mainly offered to both levels rather than just at undergraduate level.

RQ 1b: Why Is Such a Level of Corrosion Instruction Available in UAE Universities?

Open-ended questions were used to probe for reasons for the level of corrosion instruction available in the institutions. Academia respondents explained that dedicated corrosion courses were offered due to concerns over corrosion's economic and HSE impact on industries. In contrast, the reasons cited in the NRC report were broader, including employer demand, student interest and relevance to engineering careers. However, when corrosion is not taught at all in departments, the main reasons given below were consistent with the NRC report:

- other engineering topics have more priority,
- faculty not available for teaching corrosion.

RQ 2a: How Competent Are the Engineering Undergraduates/Graduates from UAE Universities in Corrosion Knowledge and Skills?

Academia and industry respondents were interviewed regarding their perceptions of engineering undergraduates/graduates as having

- sufficient fundamental knowledge of corrosion engineering (Q.34),
- sufficient understanding of the importance of corrosion in engineering design (Q.35).

The consensus was that engineering undergraduates/graduates showed insufficient fundamental knowledge of corrosion engineering due to the limited scope of corrosion in the curricula. Moreover, industry respondents highlighted the need to link theory and practice in the courses, while academia respondents cited the elective status of corrosion courses as contributing to the situation.

In contrast, respondents from University B maintained that their students had sufficient fundamental knowledge, which they attributed to the quality of their course materials: "Our students have good background knowledge because the material covered is up to date" (01.G2.R1).

Since competence involves having fundamental knowledge as well as skills in application, respondents were asked whether undergraduates/graduates understood the importance of corrosion as evidenced in their practical engineering designs. Most academia respondents held the positive view that there was sufficient understanding of the importance of corrosion in design as shown by students' awareness of the impact of corrosion and good grades and laboratory skills. However, the industry respondents were more critical in their view that graduate engineers showed superficial understanding of corrosion because of the lack of skills in application of theoretical knowledge in engineering design.

As the course status (required or elective) impacts the extent of knowledge/skills developed in the subject, views were gathered on whether corrosion engineering should be a required course in engineering degree programs. Interestingly, both academia and industry respondents were ambivalent on the issue. The position that corrosion engineering should be a required course was taken because such knowledge would be needed in the future for corrosion management in many industries. However, the opposing view that it should be anelective course was held due to the limited relevance of corrosion in some engineering fields and concern overexcessive course workload.

RQ 2b: What Resources Are Available to UAE Universities to Support Effective Corrosion Education?

The effectiveness of existing engineering curricula in corrosion is determined by both competence in corrosion knowledge/skills and availability of resources (as manpower, new knowledge through research) to achieve the desired instructional outcomes. Academia respondents were surveyed on the availability of qualified faculty and research in their departments to support corrosion education. Regarding manpower, departments in three out of the four universities would not hire faculty whose technical focus is corrosion because other topics have more priority. Interestingly, University D was the only institution where departments would hire corrosion-specialist faculty to fill newly created positions (Table 6).

Table 6: Resources: faculty recruitment

Part C	Corrosion-specialist faculty recruitment			
	University A	University B	University C	University D
Hire corrosion-specialist faculty	0%	8%	6%	55%
Do not hire corrosion-specialist faculty	100%	92%	94%	45%

The main reason cited for not recruiting a new faculty was consistent with the NRC report (Table 7). However, most of the US university respondents would hire a corrosion specialist faculty to fill a vacancy rather than a new position. This suggests that the need for qualified faculty in corrosion is being recently recognized in UAE institutions of higher education.

Table 7: Reasons for corrosion faculty recruitment

Reasons for hiring faculty*	
Fill a newly created position	77% (10)
Fill a vacancy due to a retiring or newly retired faculty member	15% (2)
Others	8% (1)
Reasons for NOT hiring faculty*	
Other topics have more priority	60% (27)
Limited department budget	7% (3)
Others	33% (15)

Regarding research, departments in three out of the four universities conducted corrosion research in varying extent, while only Universities C and D reported availability of both department research and industry partnerships in corrosion research (Table 8). In the NRC report, research funding was obtained mainly from federal agencies but the Abu Dhabi National Oil Company (ADNOC) and its operating companies were cited by University D as the main sponsors and industry partners in corrosion research.

Table 8: Resources: research

Part C	University A	University B	University C	University D
Corrosion research carried out by department	10%	0%	25%	55%
Industry research partnership available	0%	0%	13%	25%

Summary of Findings

Regarding the level of corrosion instruction, this study found that dedicated corrosion courses and engineering courses that integrated corrosion into the curricula were available in UAE universities. While the dedicated courses were offered due to concerns over corrosion's economic and HSE impact, in cases where corrosion was not taught at all, it was because other engineering topics had greater priority and a lack of qualified faculty.

Concerning the competence of engineering undergraduates/graduates in corrosion knowledge and skills, the consensus view of academia and industry respondents was that there was insufficient fundamental knowledgeof corrosion engineering due to the limited scope of corrosion in the curriculum. Moreover, the elective status and lack of effective links between theory and practice in corrosion courses contributed to the situation. Unusually, both academia and industry respondents were ambivalent on the issue of making corrosion a required course in engineering degree programs.

Besides having fundamental knowledge of corrosion, another dimension of competence is the ability to apply the theoretical knowledge in practice. Interestingly, while academia respondents held the positive view that the students displayed sufficient understanding of the importance of corrosion in engineering design, the industry respondents were more critical in their perception that graduate engineers had only superficial understandingof corrosion in real-life design contexts.

The effectiveness of engineering curricula in corrosion is determined by not only competence in corrosion knowledge/skills but also the availability of resources (as qualified faculty, development of new knowledge through research) to support corrosion education. Unfortunately, most departments would not hire corrosion-specialist faculty members because other topics have more priority. However, the aspect of research is more encouraging with two universities reporting availability of both department research and industry partnerships in corrosion research.

CONCLUSION AND RECOMMENDATIONS

Quality of Corrosion Education in the United Arab Emirates

The overarching aim of this study is to assess the quality of corrosion education in engineering programs of universities in the United Arab Emirates. The single-embedded case design adopted designated the engineering education program as the primary case and four universities formed the subunits for comparison. Between-unit comparisons of corrosion curricula characteristics were carried out in Section 4. The focus now returns to the primary case and the findings are compared to two theoretical ideal cases.

Ideal Case No.1

If corrosion is taught in the university (as dedicated and/or integrated courses), then the students/graduates would have sufficient fundamental knowledge of corrosion and skills in the application of such knowledge in practical engineering design. Also, the university would have enough resources, as qualified faculty and knowledge from research, to support corrosion education.

Ideal Case No. 2

if corrosion is not taught in the university, then the students/graduates would not have sufficient fundamental knowledge of corrosion nor skills in the application of such knowledge in practical engineering design. Also, the university would not have the resources to support corrosion education.

Table 9 shows that each subunit revealed a dimension of the phenomenon under study. It is clear that two universities were situated at opposite ends of the spectrum. University D most closely reflected Ideal Case #1whereas University A most closely represented Ideal Case #2. The remaining universities embodied variations along the scale. Within the subunits, there appeared to be discrepancies in the results that could be explained. The results for University A showed that although corrosion is not taught at all, corrosion research is carried out and this could be due to individual faculty's own effort to conduct corrosion research even though corrosion is not taught as a course. It is not unusual to observe differences between what the faculties teach and their research areas. The findings for University B showed that it offers dedicated corrosion courses but has no manpower for this activity. This project defined manpower as the availability of corrosion specialist faculty to teach corrosion courses (Table 2, Q.29). Hence, the survey asked faculty respondents if their departments would consider hiring a faculty whose technical focus is corrosion (yes/no option). Hence the data in Table 9indicated that even as University B offers dedicated corrosion courses, the departments do not intend to recruit corrosion specialist faculties to teach the courses.

Table 9: Overview of findings

	Level of corrosion instruction	Effectiveness of engineering curricula in corrosion				
	Availability	Fundamental knowledge	Skills application	Manpower	Department research	Research partnerships
Ideal case	Yes	Yes	Yes	Yes	Yes	Yes
Ideal case	No	No	No	No	No	No
University A	No	—	—	No	Yes	No
University B	Yes	Yes	Yes	No	No	No
University C	Yes	No	Yes	No	Yes	Yes
University D	Yes	No	Yes	Yes	Yes	Yes
Industry	NA	No*	No*	NA	NA	NA

*Industry respondents.

An interesting inconsistency emerged in the divergent perceptions held by academia and industry respondents regarding competence in corrosion engineering. According to the academics, the students had sufficient application skills but industry respondents held the unanimous view that graduate engineers lacked both fundamental corrosion knowledge and skills in the application of corrosion knowledge in real workplace engineering design situations. This inconsistency could be explained by the different standards used by academia and industry respondents in assessing competence. While the academia respondents measured competence as the presence of good grades and laboratory skills, industry respondents expected graduate engineers to be competent at integrating the theoretical and applied aspects of corrosion in actual practice.

In terms of the overall quality of corrosion education, this study concluded that although corrosion was taught by most engineering programs of UAE universities, there were inadequate resources to support corrosion education in the forms of corrosion-specialist faculty members as well as new knowledge gained from department research efforts and industry partnerships in corrosion research.

Recommendations for Enhancing Corrosion Knowledge and Skills

In both the surveys and interviews, respondents were asked to suggest strategies for enhancing corrosion knowledge and skills of engineering undergraduates. Three main themes emerged from the suggestions and verbatim quotes were provided to more accurately reflect the voices of the respondents and support the suggestions. This study offers the following recommendations to engineering faculty and higher education institutions for improving the knowledge base and skills of future engineers in corrosion management as well as enhancing existing corrosion instruction to better meet industry expectations.

- Set corrosion course as required course in the curriculum framework of engineering fields where corrosion is particularly relevant so that a common level of knowledge and skills could be established.
- "...engineers in mechanical, chemical and petroleum [engineering should] have to take it as a full subject" (01.G1.F3).

- Integrate corrosion theory with practical experience in engineering courses and senior design projects so that undergraduates could be actively involved in the application of corrosion theory in actual design processes.
- "Undergraduates in engineering curriculum should go on site visits with corrosion specialists … to have hands-on training on any refinery or petrochemical plant." (01.G1.F1)
- "… include corrosion in senior projects regarding means of protection" (03.G1.R5)
- "Get undergraduates to get involved in the oil industries—like studies and investigation studies regarding failures." (01.G1.A4)
- Promote awareness of corrosion impact through campus-wide workshops and activities such as Corrosion Awareness Day so as to raise students' appreciation of the need for corrosion mitigation and management.
- "… provide workshops in corrosion" (03.G3.R8)
- "… invite some people from the industry…awareness day" (03. G3.R7)
- "Bring experts to share problems (related to corrosion) with students" (03.G3.R6).

Limitations of Study and Future Research

In order to assess the quality of corrosion education in engineering programs of universities in the UAE, a qualitative case study approach was used and certain characteristics inherent in the research design resulted in two main limitations. The first limitation concerns the single-embedded case design whereby only four universities were selected as subunits within the main case. While each subunit reflected a dimension of the phenomenon that illuminated the questions under study, the results are not claimed to be generalizable to whole populations. However, the implications for enhancing quality of corrosion education derived from the findings may be extrapolated to contexts beyond the UAE.

The second limitation concerns the quality of self-reported data from interviews. Academia respondents from University B who were interviewed held the position that their students were competent (having fundamental knowledge, application skills) in corrosion engineering yet no resources (qualified faculty, research) were reported to be available in the departments to support corrosion education. It is possible that the interview questions, which required respondents to self-assess the effectiveness of their engineering curricula in corrosion, had prompted socially desirable responses that made the respondents look "good" based on fears of jeopardizing their position at the university. These main limitations are acknowledged but do not detract from the value of the findings. Instead, they indicate several avenues for future research.

Since this study showed the feasibility of the research area and validity of instruments and protocols, researchers could expand on this study with a larger sample of subunits. Besides the four universities involved, there were other universities in the UAE that matched most of the criteria (Section 3.1) except the criterion of reliable accessibility and cooperation from universities/participants. The dissemination of findings from this paper could convince other universities to participate in future studies.

Additionally, comparative studies could be conducted that examine differences in the quality of corrosion education in other Middle-East countries. Major oil/gas producing countries such as Qatar, Saudi Arabia, and Kuwait have established universities that offer extensive engineering programs. For instance, Texas A&M University in Qatar; Qatar University, King Abdullah University of Science and Technology, King Fahd University of Petroleum and Minerals, and Kuwait University. Hence, such studies could capture a more complete picture of the quality of corrosion education in a region where corrosion prevention and management are extremely crucial to its petroleum-based economies.

REFERENCES

1.   Z. Ahmad, Principles of Corrosion Engineering and Corrosion Control, Butterworth-Heinemann, Oxford, UK, 2006.

2.   P. Roberge, Corrosion Engineering: Principles and Practice, McGraw-Hill Professional, New York, NY, USA, 2008.

3.   S. Bradford, Corrosion Control, ASM International, Edmonton, Ohio, USA, 2nd edition, 2002.

4. C. J. Cron and G. A. Marsh, "Overview of economic and engineering aspects of corrosion in oil and gas production," Journal of Petroleum Technology, vol. 35, no. 7, pp. 1033–1041, 1983.

5. M. B. Kermani and D. Harrop, "The impact of corrosion on the oil and gas industry," SPE Production and Facilities, vol. 11, no. 3, pp. 186–190, 1996.

6. D. Brondel, R. Edwards, A. Hayman, D. Hill, S. Mehta, and T. Semerad, "Corrosion in the oil industry,"Oilfield Review, vol. 6, no. 2, pp. 4–18, 1994.

7. A. Al Hashem, "Corrosion in the Gulf Cooperation Council (GCC) states: statistics and figures," inProceedings of the Corrosion UAE, Abu Dhabi, UAE, 2011.

8. Business Monitor International, United Arab Emirates Oil and Gas Report Q4 2010, Business Monitor International Ltd, London, UK, 2010.

9. R. Tems and A. Al-Zahrani, "Cost of corrosion in gas sweetening and fractionation plants," inProceedings of the 61st Annual Conference and Exposition on Corrosion NACE Expo, pp. 1–12, San Diego, Calif, USA, 2006.

10. D. Rose, "Current practice: the teaching of corrosion in colleges and universities," in Materials Forum 2007: Corrosion Education for the 21st Century, pp. 8–12, Washington, DC, USA, 2007.

11. National Research Council, Assessment of Corrosion Education, The National Academy of Sciences, Washington, DC, USA, 2009.

12. R. Yin, Case Study Research: Design and Methods, Sage, Beverly Hills, Calif, USA, 2nd edition, 1994.

13. R. Yin, Case Study Research: Design and Methods, vol. 5, Sage, Thousand Oaks, Calif, USA, 3rd edition, 2003.

14. M. Patton, Qualitative Research and Evaluation Methods, Sage, Thousand Oaks, Calif, USA, 3rd edition, 2002.

15. I. Holloway and S. Wheeler, "Ethical issues in qualitative nursing research," Nursing Ethics, vol. 2, no. 3, pp. 223–232, 1995.

Chapter 8

A Review on Nanofluids: Preparation, Stability Mechanisms, and Applications

Wei Yu and Huaqing Xie

School of Urban Development and Environmental Engineering, Shanghai Second Polytechnic University, Shanghai 201209, China

ABSTRACT

Nanofluids, the fluid suspensions of nanomaterials, have shown many interesting properties, and the distinctive features offer unprecedented potential for many applications. This paper summarizes the recent progress on the study of nanofluids, such as the preparation methods, the evaluation methods for the stability of nanofluids, and the ways to enhance the stability for nanofluids, the stability mechanisms of nanofluids, and presents the broad range of current and future applications in various fields including energy and mechanical and biomedical fields. At last, the paper identifies the opportunities for future research.

INTRODUCTION

Nanofluids are a new class of fluids engineered by dispersing nanometer-sized materials (nanoparticles, nanofibers, nanotubes, nanowires, nanorods, nanosheet, or droplets) in base fluids. In other words, nanofluids are nanoscale colloidal suspensions containing condensed nanomaterials. They are two-phase systems with one phase (solid phase) in another (liquid phase). Nanofluids have been found to possess enhanced thermophysical properties such as thermal conductivity, thermal diffusivity, viscosity, and convective heat transfer coefficients compared to those of base fluids like oil or water. It has demonstrated great potential applications in many fields.

For a two-phase system, there are some important issues we have to face. One of the most important issues is the stability of nanofluids, and it remains a big challenge to achieve desired stability of nanofluids. In this paper, we will review the new progress in the methods for preparing stable nanofluids and summarize the stability mechanisms.

In recent years, nanofluids have attracted more and more attention. The main driving force for nanofluids research lies in a wide range of applications. Although some review articles involving the progress of nanofluid investigation were published in the past several years [1–6], most of the reviews are concerned of the experimental and theoretical studies of the thermophysical properties or the convective heat transfer of nanofluids. The purpose of this paper will focuses on the new preparation methods and stability mechanisms, especially the new application trends for nanofluids in addition to the heat transfer properties of nanofluids. We will try to find some challenging issues that need to be solved for future research based on the review on these aspects of nanofluids.

PREPARATION METHODS FOR NANOFLUIDS

Two-Step Method

Two-step method is the most widely used method for preparing nanofluids. Nanoparticles, nanofibers, nanotubes, or other nanomaterials used in this method are first produced as dry powders by chemical or physical methods. Then, the nanosized powder will be dispersed into a fluid in the second processing step with the help of intensive magnetic force agitation, ultrasonic agitation, high-shear mixing, homogenizing, and ball milling. Two-step method is the most economic method to produce nanofluids in large scale, because nanopowder synthesis techniques have already been scaled up to industrial production levels. Due to the high surface area and surface activity, nanoparticles have the tendency to aggregate. The important technique to enhance the stability of nanoparticles in fluids is the use of surfactants. However, the functionality of the surfactants under high temperature is also a big concern, especially for high-temperature applications.

Due to the difficulty in preparing stable nanofluids by two-step method, several advanced techniques are developed to produce nanofluids, including one-step method. In the following part, we will introduce one-step method in detail.

One-Step Method

To reduce the agglomeration of nanoparticles, Eastman et al. developed a one-step physical vapor condensation method to prepare Cu/ethylene glycol nanofluids [7]. The one-step process consists of simultaneously making and dispersing the particles in the fluid. In this method, the processes of drying, storage, transportation, and dispersion of nanoparticles are avoided, so the agglomeration of nanoparticles is minimized, and the stability of fluids is increased [5]. The one-step processes can prepare uniformly dispersed nanoparticles, and the particles can be stably suspended in the base fluid. The vacuum-SANSS (submerged arc nanoparticle synthesis system) is another efficient method to prepare nanofluids using different dielectric liquids [8, 9]. The different morphologies are mainly influenced and determined by various thermal conductivity properties of the dielectric liquids. The nanoparticles prepared exhibit needle-like, polygonal, square, and circular morphological shapes. The method avoids the undesired particle aggregation fairly well.

One-step physical method cannot synthesize nanofluids in large scale, and the cost is also high, so the one-step chemical method is developing rapidly. Zhu et al. presented a novel one-step chemical method for preparing copper nanofluids by reducing $CuSO_4 \cdot 5 H_2O$ with $NaH_2PO_2 \cdot H_2O$ in ethylene glycol under microwave irradiation [10]. Well-dispersed and stably suspended copper nanofluids were obtained. Mineral oil-based nanofluids containing silver nanoparticles with a narrow-size distribution were also prepared by this method [11]. The particles could be stabilized by Korantin, which coordinated to the silver particle surfaces via two oxygen atoms forming a dense layer around the particles. The silver nanoparticle suspensions were stable for about 1 month. Stable ethanol-based nanofluids containing silver nanoparticles could be prepared by microwave-assisted one-step method [12]. In the method, polyvinyl pyrrolidone (PVP) was employed as the stabilizer of colloidal silver and reducing agent for silver in solution. The cationic surfactant octadecylamine (ODA) is also an efficient phase-transfer agent to synthesize silver colloids [13]. The phase transfer of the silver nanoparticles arises due to coupling of the silver nanoparticles with the ODA molecules present in organic phase via either coordination bond formation or weak covalent interaction. Phase transfer method has been developed for preparing homogeneous and stable graphene oxide colloids. Graphene oxide nanosheets (GONs) were successfully transferred from water to n-octane after modification by oleylamine, and the schematic illustration of the phase transfer process is shown in Figure 1 [14].

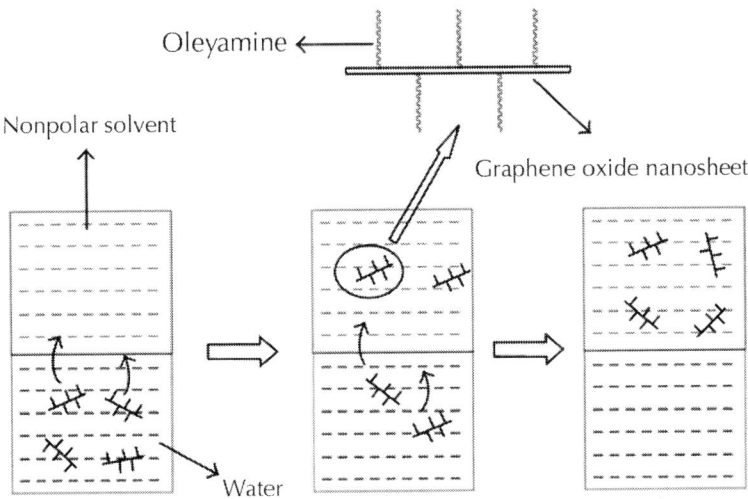

Figure 1: Schematic illustration of the phase transfer process.

However, there are some disadvantages for one-step method. The most important one is that the residual reactants are left in the nanofluids due to incomplete reaction or stabilization. It is difficult to elucidate the nanoparticle effect without eliminating this impurity effect.

Other Novel Methods

Wei et al. developed a continuous-flow microfluidic microreactor to synthesize copper nanofluids. By this method, copper nanofluids can be continuously synthesized, and their microstructure and properties can be varied by adjusting parameters such as reactant concentration, flow rate, and additive. CuO nanofluids with high solid volume fraction (up to 10 vol %) can be synthesized through a novel precursor transformation method with the help of ultrasonic and microwave irradiation [15]. The precursor Cu $(OH)_2$ is completely transformed to CuO nanoparticle in water under microwave irradiation. The ammonium citrate prevents the growth and aggregation of nanoparticles, resulting in a stable CuO aqueous nanofluid with higher thermal conductivity than those prepared by other dispersing methods. Phase-transfer method is also a facile way to obtain monodisperse noble metal colloids [16]. In a water-cyclohexane two-phase system, aqueous formaldehyde is transferred to cyclohexane phase via reaction with dodecylamine to form reductive intermediates in cyclohexane. The intermediates are capable of reducing silver or gold ions in aqueous solution to form dodecyl amine-protected silver and gold nanoparticles in cyclohexane solution at room temperature. Feng et al. used the aqueous organic phase-transfer method for preparing gold, silver, and platinum nanoparticles on the basis of the decrease of the PVP's solubility in water with the temperature increase [17]. Phase-transfer method is also applied for preparing stable kerosene-based Fe_3O_4 nanofluids. Oleic acid is successfully grafted onto the surface of Fe_3O_4 nanoparticles by chemisorbed mode, which lets Fe_3O_4 nanoparticles have good compatibility with kerosene [18]. The Fe_3O_4 nanofluids prepared by phase-transfer method do not show the previously reported "time dependence of the thermal conductivity characteristic". The preparation of nanofluids with controllable microstructure is one of the key issues. It is well known that the properties of nanofluids strongly depend on the structure and shape of nanomaterials. The recent research shows that nanofluids synthesized by chemical solution method have both higher conductivity enhancement and better stability than those produced by the other methods [19]. This method is distinguished from the others by its controllability. The nanofluid microstructure can be varied and manipulated by adjusting synthesis parameters such as temperature, acidity, ultrasonic and microwave irradiation, types and concentrations of reactants and additives, and the order in which the additives are added to the solution.

THE STABILITY OF NANOFLUID

The agglomeration of nanoparticles results in not only the settlement and clogging of microchannels but also the decreasing of thermal conductivity of nanofluids. So, the investigation on stability is also a key issue that influences the properties of nanofluids for application, and it is necessary to study and analyze influencing factors to the dispersion stability of nanofluids. This section will contain (a) the stability evaluation methods for nanofluids, (b) the ways to enhance the stability of nanofluids, and (c) the stability mechanisms of nanofluids.

The Stability Evaluation Methods for Nanofluids

Sedimentation and Centrifugation Methods

Many methods have been developed to evaluate the stability of nanofluids. The simplest method is sedimentation method [20, 21]. The sediment weight or the sediment volume of nanoparticles in a nanofluid under an external force field is an indication of the stability of the characterized nanofluid. The variation of concentration or particle size of supernatant particle with sediment time can be obtained by special apparatus [5]. The nanofluids are considered to be stable when the concentration or particle size of supernatant particles keeps constant. Sedimentation photograph of nanofluids in test tubes taken by a camera is also a usual method for observing the stability of nanofluids [5]. Zhu et al. used a sedimentation balance method to measure the stability of the graphite suspension [22]. The tray of sedimentation balance immerged in the fresh graphite suspension. The weight of sediment nanoparticles during a certain period was measured. The suspension fraction of graphite nanoparticles at a certain time could be calculated. For the sedimentation method, long period for observation is the defect. Therefore, centrifugation method is developed to evaluate the stability of nanofluids. Singh et al. applied the centrifugation method to observe the stability of silver nanofluids prepared by the microwave synthesis in ethanol by reduction of $AgNO_3$ with PVP as stabilizing agent [12]. It has been found that the obtained nanofluids are stable for more than 1 month in the stationary state and more than 10 h under centrifugation at 3,000 rpm without sedimentation. Excellent stability of the obtained nanofluid is due to the protective role of PVP, as it retards the growth and agglomeration of nanoparticles by steric effect. Li prepared the aqueous polyaniline colloids and used the centrifugation method to evaluate the stability of the colloids [23]. Electrostatic repulsive forces between nanofibers enabled the long-term stability of the colloids.

Zeta Potential Analysis

Zeta potential is electric potential in the interfacial double layer at the location of the slipping plane versus a point in the bulk fluid away from the interface, and it shows the potential difference between the dispersion medium and the stationary layer of fluid attached to the dispersed particle. The significance of zeta potential is that its value can be related to the stability of colloidal dispersions. So, colloids with high zeta potential (negative or positive) are electrically stabilized, while colloids with low zeta potentials tend to coagulate or flocculate. In general, a value of 25 mV (positive or negative) can be taken as the arbitrary value that separates low-charged surfaces from highly charged surfaces. The colloids with zeta potential from 40 to 60 mV are believed to be good stable, and those with more than 60 mV have excellent stability. Kim et al. prepared Au nanofluids with an outstanding stability even after 1 month although no dispersants were observed [24]. The stability is due to a large negative zeta potential of Au nanoparticles in water. The influence of pH and sodium dodecylbenzene sulfonate (SDBS) on the stability of two water-based nanofluids was studied [25], and zeta potential analysis was an important technique to evaluate the stability. Zhu et al. [26] measured the zeta potential of Al_2O_3-H_2O nanofluids under different pH values and different SDBS concentration. The Derjaguin-Laudau-Verwey-Overbeek (DLVO) theory was used to calculate attractive and repulsive potentials. Cationic gemini surfactant as stabilizer was used to prepare stable water-based nanofluids containing MWNTs [27]. Zeta potential measurements were employed to study the absorption mechanisms of the surfactants on the MWNT surfaces with the help of Fourier transformation infrared spectra.

Spectral Absorbency Analysis

Spectral absorbency analysis is another efficient way to evaluate the stability of nanofluids. In general, there is a linear relationship between the absorbency intensity and the concentration of nanoparticles in fluid. Huang et al. evaluated the dispersion characteristics of alumina and copper suspensions using the conventional sedimentation method with the help of absorbency analysis by using a spectrophotometer after the suspensions deposited for 24 h [28]. The stability investigation of colloidal FePt nanoparticle systems was done via spectrophotometer analysis [29]. The sedimentation kinetics could also be determined by examining the absorbency of particle in solution [26].

If the nanomaterials dispersed in fluids have characteristic absorption bands in the wavelength 190–1100 nm, it is an easy and reliable method to evaluate the stability of nanofluids using UV-vis spectral analysis. The variation of supernatant particle concentration of nanofluids with sediment time can be obtained by the measurement of absorption of nanofluids, because there is a linear relation between the supernatant nanoparticle concentration and the absorbance of suspended particles. The outstanding advantage comparing to other methods is that UV-vis spectral analysis can present the quantitative concentration of nanofluids. Hwang et al. [30] studied the stability of nanofluids with the UV-vis spectrophotometer. It was believed that the stability of nanofluids was strongly affected by the characteristics of the suspended particles and the base fluid such as particle morphology. Moreover, the addition of a surfactant could improve the stability of the suspensions. The relative stability of MWNT nanofluids [27] could be estimated by measuring the UV-vis absorption of the MWNT nanofluids at different sediment times. From the above relation between MWNT concentration and its UV-vis absorbance value, the concentration of the MWNT nanofluids at different sediment times could be obtained. The above three methods can be united to investigate the stability of nanofluids. For example, Li et al. evaluated the dispersion behavior of the aqueous copper nanosuspensions under different pH values, different dispersant type, and concentration by the method of zeta potential, absorbency, and sedimentation photographs [21].

The Ways to Enhance the Stability of Nanofluids

Surfactants Used in Nanofluids

Surfactants used in nanofluids are also called dispersants. Adding dispersants in the two-phase systems is an easy and economic method to enhance the stability of nanofluids. Dispersants can markedly affect the surface characteristics of a system in small quantity. Dispersants consists of a hydrophobic tail portion, usually a long-chain hydrocarbon, and a hydrophilic polar head group. Dispersants are employed to increase the contact of two materials, sometimes known as wettability. In a two-phase system, a dispersant tends to locate at the interface of the two phases, where it introduces a degree of continuity between the nanoparticles and fluids. According to the composition of the head, surfactants are divided into four classes: nonionic surfactants without charge groups in its head (include polyethylene oxide, alcohols, and other polar groups), anionic surfactants with negatively charged head groups (anionic head groups include long-chain fatty acids, sulfosuccinates, alkyl sulfates, phosphates, and sulfonates), cationic surfactants with positively charged head groups (cationic surfactants may be protonated long-chain amines and long-chain quaternary ammonium compounds), and amphoteric surfactants with zwitterionic head groups (charge depends on pH. The class of amphoteric surfactants is represented by betaines and certain lecithins). How to select suitable dispersants is a key issue. In general, when the base fluid of nanofluids is polar solvent, we should select water-soluble surfactants; otherwise, we will select oil-soluble ones. For nonionic surfactants, we can evaluate the solubility through the term hydrophilic/lipophilic balance (HLB) value. The lower the HLB number, the more oil-soluble the surfactants, and in turn, the higher the HLB number, the more water-soluble the surfactants is. The HLB value can be obtained easily by many handbooks. Although surfactant addition is an effective way to enhance the dispersibility of nanoparticles, surfactants might cause several problems [31]. For example, the addition of surfactants may contaminate the heat transfer media. Surfactants may produce foams when heating, while heating and cooling are routine processes in heat exchange systems. Furthermore, surfactant molecules attaching on the surfaces of nanoparticles may enlarge the thermal resistance between the nanoparticles and the base fluid, which may limit the enhancement of the effective thermal conductivity.

Surface Modification Techniques: Surfactant-Free Method

Use of functionalized nanoparticles is a promising approach to achieve long-term stability of nanofluid. It represents the surfactant-free technique. Yang and Liu presented a work on the synthesis of functionalized silica (SiO_2) nanoparticles by grafting silanes directly to the surface of silica nanoparticles in original nanoparticle solutions [32]. One of the unique characteristics of the nanofluids was that no deposition layer formed on the heated surface after a pool boiling process. Hwang et al. introduced hydrophilic functional groups on the surface of the nanotubes by mechanochemical reaction [30]. The prepared nanofluids, with no contamination to medium, good fluidity, low viscosity, high stability, and high thermal conductivity, would have potential applications as coolants in advanced thermal systems. A wet mechanochemical reaction was applied to prepare surfactant-free nanofluids containing double- and single-walled CNTs. Results from the infrared spectrum and zeta potential measurements showed that the hydroxyl groups had been introduced onto the treated CNT surfaces [33]. The chemical modification to functionalize the surface of carbon nanotubes is a common method to enhance the stability of carbon nanotubes in solvents. Here, we present a review about the surface modification of carbon nanotubes [34]. Plasma treatment was used to modify the surface characteristics of diamond nanoparticles [35]. Through plasma treatment using gas mixtures of methane and oxygen, various polar groups were imparted on the surface of the diamond nanoparticles, improving their dispersion property in water. A stable dispersion of titania nanoparticles in an organic solvent of diethylene glycol dimethylether (diglyme) was successfully prepared using a ball milling process [36]. In order to enhance dispersion stability of the solution, surface modification of dispersed titania particles was carried out during the centrifugal bead mill process. Surface modification was utilized with silane coupling agents, (3-acryl-oxypropyl) trimethoxysilane and trimethoxypropylsilane. Zinc oxide nanoparticles could be modified by polymethacrylic acid (PMAA) in aqueous system [37]. The hydroxyl groups of nano-ZnO particle surface could interact with carboxyl groups of PMAA and form poly (zinc methacrylate) complex on the surface of nano-ZnO. PMAA enhanced the dispersibility of nano-ZnO particles in water. The modification did not alter the crystalline structure of the ZnO nanoparticles.

Stability Mechanisms of Nanofluids

Particles in dispersion may adhere together and form aggregates of increasing size which may settle out due to gravity. Stability means that the particles do not aggregate at a significant rate. The rate of aggregation is in general determined by the frequency of collisions and the probability of cohesion during collision. Derjaguin, Verway, Landau, and Overbeek (DVLO) developed a theory which dealt with colloidal stability [38, 39]. DLVO theory suggests that the stability of a particle in solution is determined by the sum of van der Waals attractive and electrical double layer repulsive forces that exist between particles as they approach each other due to the Brownian motion they are undergoing. If the attractive force is larger than the repulsive force, the two particles will collide, and the suspension is not stable. If the particles have a sufficient high repulsion, the suspensions will exist in stable state. For stable nanofluids or colloids, the repulsive forces between particles must be dominant. According to the types of repulsion, the fundamental mechanisms that affect colloidal stability are divided into two kinds, one is steric repulsion, and another is electrostatic (charge) repulsion, shown in Figure2. For steric stabilization, polymers are always involved into the suspension system, and they will adsorb onto the particles surface, producing an additional steric repulsive force. For example, Zinc oxide nanoparticles modified by PMAA have good compatibility with polar solvents [37]. Silver nanofluids are very stable due to the protective role of PVP, as it retards the growth and agglomeration of nanoparticles by steric effect. PVP is an efficient agent to improve the stability of graphite suspension [22]. The steric effect of polymer dispersant is determined by the concentration of the dispersant. If the PVP concentration is low, the surface of the graphite particles is gradually coated by PVP molecules with the increase of PVP. Kamiya et al. studied the effect of polymer dispersant structure on electrostatic interaction and dense alumina suspension behavior [40]. An optimum hydrophilic to hydrophobic group ratio was obtained from the maximum repulsive force and minimum viscosity. For electrostatic stabilization, surface charge will be developed through one or more of the following mechanisms: (1) preferential adsorption of ions, (2) dissociation of surface charged species, (3) isomorphic substitution of ionsm, (4) accumulation or depletion of electrons at the surface, and (5) physical adsorption of charged species onto the surface.

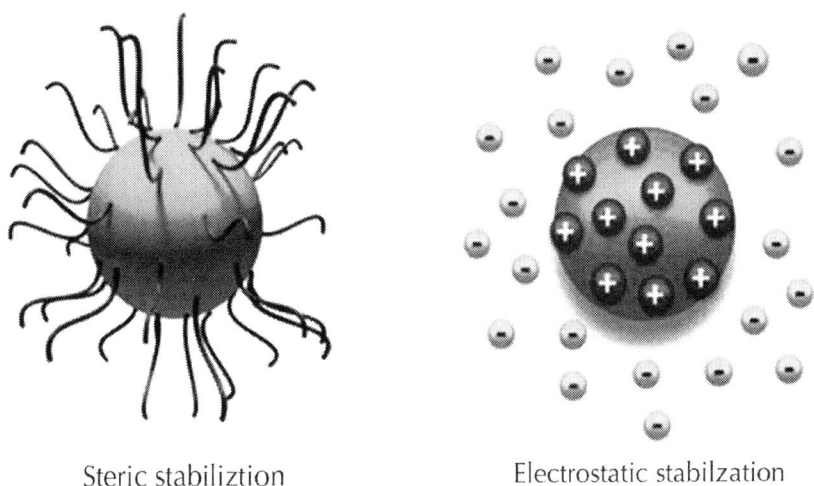

Steric stabiliztion   Electrostatic stabilzation

Figure 2: Types of colloidal stabilization.

APPLICATION OF NANOFLUIDS

Heat Transfer Intensification

Since the origination of the nanofluid concept about a decade ago, the potentials of nanofluids in heat transfer applications have attracted more and more attention. Up to now, there are some review papers which present overviews of various aspects of nanofluids [1, 3–6, 41–46], including preparation and characterization, techniques for the measurements of thermal conductivity, theory and model, thermophysical properties, and convective heat transfer. Our group studied the thermal conductivities of ethylene glycol- (EG-) based nanofluids containing oxides including MgO, TiO_2, ZnO, Al_2O_3, and SiO_2 nanoparticles [47], and the results (Table1) demonstrated that MgO-EG nanofluid was found to have superior features with the highest thermal conductivity and lowest viscosity. In this part, we will summarize the applications of nanofluids in heat transfer enhancement.

Table 1: Properties of oxides and their nanofluids

	Thermal conductivity* W/(m·K)	Density (g/cm³)	Crystalline	Viscosity (Cp) with 5.0 vol. % 30°C	Thermal conductivity enhancement of nanofluids (%) with 5.0 vol. %
MgO	48.4	2.9	Cubic	17.4	40.6
TiO₂	8.4	4.1	Anatase	31.2	27.2
ZnO	13.0	5.6	Wurtzite	129.2	26.8
Al₂O₃	36.0	3.6		28.2	28.2
SiO₂	10.4	2.6	noncrystalline	31.5	25.3

*Thermal conductivities of the oxides are for the corresponding bulk materials.

Electronic Applications

Due to higher density of chips, design of electronic components with more compact makes heat dissipation more difficult. Advanced electronic devices face thermal management challenges from the high level of heat generation and the reduction of available surface area for heat removal. So, the reliable thermal management system is vital for the smooth operation of the advanced electronic devices. In general, there are two approaches to improve the heat removal for electronic equipment. One is to find an optimum geometry of cooling devices; another is to increase the heat transfer capacity. Nanofluids with higher thermal conductivities are predicated convective heat transfer coefficients compared to those of base fluids. Recent researches illustrated that nanofluids could increase the heat transfer coefficient by increasing the thermal conductivity of a coolant. Jang and Choi designed a new cooler, combined microchannel heat sink with nanofluids [48]. Higher cooling performance was obtained when compared to the device using pure water as working medium. Nanofluids reduced both the thermal resistance and the temperature difference between the heated microchannel wall and the coolant. A combined microchannel heat sink with nanofluids had the potential as the next-generation cooling devices for removing ultrahigh heat flux. Nguyen et al. designed a closed liquid-circuit to investigate the heat transfer enhancement of a liquid cooling system by replacing the base fluid (distilled water) with a nanofluid composed of distilled water and Al_2O_3 nanoparticles at various concentrations [49]. Measured data have clearly shown that the inclusion of nanoparticles within the distilled water has produced a considerable enhancement in convective heat transfer coefficient of the cooling block. With particle loading 4.5 vol%, the enhancement is up to 23% with respect to that of the base fluid. It has also been observed that an augmentation of particle concentration has produced a clear decrease of the junction temperature between the heated component and the cooling block. Silicon microchannel heat sink performance using nanofluids containing Cu nanoparticles was analyzed [50]. It was found that nanofluids could enhance the performance as compared with that using pure water as the coolant. The enhancement was due to the increase in thermal conductivity of coolant and the nanoparticle thermal dispersion effect. The other advantage was that there was no extra pressure drop, since the nanoparticle was small, and particle volume fraction was low.

The thermal requirements on the personal computer become much stricter with the increase in thermal dissipation of CPU. One of the solutions is the use of heat pipes. Nanofluids, employed as working medium for conventional heat pipe, have shown higher thermal performances, having the potential as a substitute for conventional water in heat pipe. At a same charge volume, there is a significant reduction in thermal resistance of heat pipe with nanofluid containing gold nanoparticles as compared with water [51]. The measured results also show that the thermal resistance of a vertical meshed heat pipe varies with the size of gold nanoparticles. The suspended nanoparticles tend to bombard the vapor bubble during the bubble formation. Therefore, it is expected that the nucleation size of vapor bubble is much smaller for fluid with suspended nanoparticles than that without them. This may be the major reason for reducing the thermal resistance of heat pipe. Chen et al. studied the effect of a nanofluid on flat heat pipe (FHP) thermal performance [52], using silver nanofluid as the working fluid. The temperature difference and the thermal resistance of the FHP with the silver nanoparticle solution were lower than those with pure water. The plausible reasons for enhancement of the thermal performance of the FHP using the nanofluid can be explained by the critical heat flux enhancement by higher wettability and the reduction of the boiling limit. Nanofluid oscillating heat pipe with ultrahigh-performance was developed by Ma et al. [53]. They combined nanofluids with thermally excited oscillating motion in an oscillating heat pipe, and heat transport capability significantly increased. For example, at the input power of 80.0 W, diamond nanofluid could reduce the temperature difference between the evaporator and the condenser from 40.9 to 24.3°C. This study would accelerate the development of a highly efficient cooling device for ultrahigh-heat-flux electronic systems. The thermal performance investigation of heat pipe indicated that nanofluids containing silver or titanium nanoparticles could be used as an efficient cooling fluid for devices with high energy density. For a silver nanofluid, the temperature difference decreased 0.56–0.65 compared to water at an input power of 30–50 W [54]. For the heat pipe with titanium nanoparticles at a volume concentration of 0.10%, the thermal efficiency is 10.60% higher than that with the based working fluid [55]. These positive results are promoting the continued research and development of nanofluids for such applications.

Transportation

Nanofluids have great potentials to improve automotive and heavy-duty engine cooling rates by increasing the efficiency, lowering the weight and reducing the complexity of thermal management systems. The improved cooling rates for automotive and truck engines can be used to remove more heat from higher horsepower engines with the same size of cooling system. Alternatively, it is beneficial to design more compact cooling system with smaller and lighter radiators. It is, in turn, beneficial the high performance and high fuel economy of car and truck. Ethylene glycol-based nanofluids have attracted much attention in the application as engine coolant [56–58] due to the low-pressure operation compared with a 50/50 mixture of ethylene glycol and water, which is the nearly universally used automotive coolant. The nanofluids has a high boiling point, and it can be used to increase the normal coolant operating temperature and then reject more heat through the existing coolant system [59]. Kole et al. prepared car engine coolant (Al_2O_3 nanofluid) using a standard car engine coolant (HP KOOLGARD) as the base fluid [60] and studied the thermal conductivity and viscosity of the coolant. The prepared nanofluid, containing only 3.5% volume fraction of Al_2O_3 nanoparticles, displayed a fairly higher thermal conductivity than the base fluid, and a maximum enhancement of 10.41% was observed at room temperature. Tzeng et al. [61] applied nanofluids to the cooling of automatic transmissions. The experimental platform was the transmission of a four-wheel drive vehicle. The used nanofluids were prepared by dispersing CuO and Al_2O_3 nanoparticles into engine transmission oil. The results showed that CuO nanofluids produced the lower transmission temperatures both at high and low rotating speeds. From the thermal performance viewpoint, the use of nanofluid in the transmission has a clear advantage.

The researchers of Argonne National Laboratory have assessed the applications of nanofluids for transportation [62]. The use of high-thermal conductive nanofluids in radiators can lead to a reduction in the frontal area of the radiator up to 10%. The fuel saving is up to 5% due to the reduction in aerodynamic drag. It opens the door for new aerodynamic automotive designs that reduce emissions by lowering drag. The application of nanofluids also contributed to a reduction of friction and wear, reducing parasitic losses, operation of components such as pumps and compressors, and subsequently leading to more than 6% fuel savings. In fact, nanofluids not only enhance the efficiency and economic performance of car engine, but also will greatly influence the structure design of automotives. For example, the engine radiator cooled by a nanofluid will be smaller and lighter. It can be placed elsewhere in the vehicle, allowing for the redesign of a far more aerodynamic chassis. By reducing the size and changing the location of the radiator, a reduction in weight and wind resistance could enable greater fuel efficiency and subsequently lower exhaust emissions. Computer simulations from the US department of energy's office of vehicle technology showed that nanofluid coolants could reduce the size of truck radiators by 5%. This would result in a 2.5% fuel saving at highway speeds.

The practical applications are on the road. In USA, car manufacturers GM and Ford are running their own research programs on nanofluid applications. A 8.3 million FP7 project, named NanoHex (Nanofluid Heat Exchange), began to run. It involved 12 organizations from Europe and Israel ranging from Universities to SMEs and major companies. NanoHex is overcoming the technological challenges faced in development and application of reliable and safe nanofluids for more sophisticated, energy efficient, and environmentally friendly products and services [63].

Industrial Cooling Applications

The application of nanofluids in industrial cooling will result in great energy savings and emissions reductions. For US industry, the replacement of cooling and heating water with nanofluids has the potential to conserve 1 trillion Btu of energy [41, 64]. For the US electric power industry, using nanofluids in closed loop cooling cycles could save about 10–30 trillion Btu per year (equivalent to the annual energy consumption of about 50,000–150,000 households). The associated emissions reductions would be approximately 5.6 million metric tons of carbon dioxide, 8,600 metric tons of nitrogen oxides, and 21,000 metric tons of sulfur dioxide [65].

Experiments were performed using a flow-loop apparatus to explore the performance of polyalphaolefin nanofluids containing exfoliated graphite nanoparticle fibers in cooling [66]. It was observed that the specific heat of nanofluids was found to be 50% higher for nanofluids compared with polyalphaolefin, and it increased with temperature. The thermal diffusivity was found to be 4 times higher for nanofluids. The convective heat transfer was enhanced by 10% using nanofluids compared with using polyalphaolefin. Ma et al. proposed the concept of nanoliquid-metal fluid, aiming to establish an engineering route to make the highest conductive coolant with about several dozen times larger thermal conductivity than that of water [45]. The liquid metal with low melting point is expected to be an idealistic base fluid for making superconductive solution, which may lead to the ultimate coolant in a wide variety of heat transfer enhancement area. The thermal conductivity of the liquid-metal fluid can be enhanced through the addition of more conductive nanoparticles.

Heating Buildings and Reducing Pollution

Nanofluids can be applied in the building heating systems. Kulkarni et al. evaluated how they perform heating buildings in cold regions [67]. In cold regions, it is a common practice to use ethylene or propylene glycol mixed with water in different proportions as a heat transfer fluid. So, 60:40 ethylene glcol/water (by weight) was selected as the base fluid. The results showed that using nanofluids in heat exchangers could reduce volumetric and mass flow rates, resulting in an overall pumping power savings. Nanofluids necessitate smaller heating systems, which are capable of delivering the same amount of thermal energy as larger heating systems but are less expensive. This lowers the initial equipment cost excluding nanofluid cost. This will also reduce environmental pollutants, because smaller heating units use less power, and the heat transfer unit has less liquid and material waste to discard at the end of its life cycle.

Nuclear Systems Cooling

The Massachusetts Institute of Technology has established an interdisciplinary center for nanofluid technology for the nuclear energy industry. The researchers are exploring the nuclear applications of nanofluids, specifically the following three [68]: (1) main reactor coolant for pressurized water reactors (PWRs). It could enable significant power uprates in current and future PWRs, thus enhancing their economic performance. Specifically, the use of nanofluids with at least 32% higher critical heat flux (CHF) could enable a 20% power density uprate in current plants without changing the fuel assembly design and without reducing the margin to CHF; (2) coolant for the emergency core cooling systems (ECCSs) of both PWRs and boiling water reactors. The use of a nanofluid in the ECCS accumulators and safety injection can increase the peak-cladding-temperature margins (in the nominal-power core) or maintain them in uprated cores if the nanofluid has a higher post-CHF heat transfer rate; (3) coolant for in-vessel retention of the molten core during severe accidents in high-power-density light water reactors. It can increase the margin to vessel breach by 40% during severe accidents in high-power density systems such as Westinghouse APR1000 and the Korean APR1400. While there exist several significant gaps, including the nanofluid thermal-hydraulic performance at prototypical reactor conditions and the compatibility of the nanofluid chemistry with the reactor materials. Much work should be done to overcome these gaps before any applications can be implemented in a nuclear power plant.

Space and Defense

Due to the restriction of space, energy, and weight in space station and aircraft, there is a strong demand for high efficient cooling system with smaller size. You et al. [69] and Vassalo et al. [70] have reported order of magnitude increases in the critical heat flux in pool boiling with nanofluids compared to the base fluid alone. Further research of nanofluids will lead to the development of next generation of cooling devices that incorporate nanofluids for ultrahigh-heat-flux electronic systems, presenting the possibility of raising chip power in electronic components or simplifying cooling requirements for space applications. A number of military devices and systems require high-heat flux cooling to the level of tens of MW/m^2. At this level, the cooling of military devices and system is vital for the reliable operation. Nanofluids with high critical heat fluxes have the potential to provide the required cooling in such applications as well as in other military systems, including military vehicles, submarines, and high-power laser diodes. Therefore, nanofluids have wide application in space and defense fields, where power density is very high and the components should be smaller and weight less.

Mass Transfer Enhancement

Several researches have studied the mass transfer enhancement of nanofluids. Kim ET al.initially examined the effect of nanoparticles on the bubble type absorption for NH_3/H_2O absorption system [71]. The addition of nanoparticles enhances the absorption performance up to 3.21 times. Then, they visualized the bubble behavior during the NH_3/H_2O absorption process and studied the effect of nanoparticles and surfactants on the absorption characteristics [72]. The results show that the addition of surfactants and nanoparticles improved the absorption performance up to 5.32 times. The addition of both surfactants and nanoparticles enhanced significantly the absorption performance during the ammonia bubble absorption process. The theoretical investigations of thermodiffusion and diffusionthermo on convective instabilities in binary nanofluids for absorption application were conducted. Mass diffusion is induced by thermal gradient. Diffusionthermo implies that heat transfer is induced by concentration gradient [73]. Ma et al. studied the mass transfer process of absorption using CNTs-ammonia nanofluids as the working medium [74, 75]. The absorption rates of the CNTs-ammonia binary nanofluids were higher than those of ammonia solution without CNTs. The effective absorption ratio of the CNTs-ammonia binary nanofluids increased with the initial concentration of ammonia and the mass fraction of CNTs. Komati et al. studied CO_2 absorption into amine solutions, and the addition of ferrofluids increased the mass transfer coefficient in gas/liquid mass transfer [76], and the enhancement extent depended on the amount of ferrofluid added. The enhancement in mass transfer coefficient was 92.8% for a volume fraction of the fluid of about 50% (solid magnetite volume fraction of about 0.39%). The research about the influence of Al_2O_3 nanofluid on the falling film absorption with ammonia water showed that the sorts of nanoparticles and surfactants in the nanofluid and the concentration of ammonia in the basefluid were the key parameters influencing the absorption effect of ammonia [77].

So far, the mechanism leading to mass transfer enhancement is still unclear. The existing research work on the mass transfer in nanofluids is not enough. Much experimental and simulation work should be carried out to clarify some important influencing factors.

Energy Applications

For energy applications of nanofluids, two remarkable properties of nanofluids are utilized, one is the higher thermal conductivities of nanofluids, enhancing the heat transfer, and another is the absorption properties of nanofluids.

Energy Storage

The temporal difference of energy source and energy needs made necessary the development of storage system. The storage of thermal energy in the form of sensible and latent heat has become an important aspect of energy management with the emphasis on efficient use and conservation of the waste heat and solar energy in industry and buildings [78]. Latent heat storage is one of the most efficient ways of storing thermal energy. Wu et al. evaluated the potential of Al_2O_3-H_2O nanofluids as a new phase change material (PCM) for the thermal energy storage of cooling systems. The thermal response test showed the addition of Al_2O_3 nanoparticles remarkably decreased the supercooling degree of water, advanced the beginning freezing time, and reduced the total freezing time. Only adding 0.2 wt% Al_2O_3 nanoparticles, the total freezing time of Al_2O_3-H_2O nanofluids could be reduced by 20.5%. Liu et al. prepared a new sort of nanofluid phase change materials (PCMs) by suspending small amount of TiO_2 nanoparticles in saturated $BaCl_2$ aqueous solution [79]. The nanofluids PCMs possessed remarkably high thermal conductivities compared to the base material. The cool storage/supply rate and the cool storage/supply capacity all increased greatly than those of $BaCl_2$ aqueous solution without added nanoparticles. The higher thermal performances of nanofluids PCMs indicate that they have a potential for substituting conventional PCMs in cool storage applications. Copper nanoparticles are efficient additives to improve the heating and cooling rates of PCMs [80]. For composites with 1 wt % copper nanoparticle, the heating and cooling times could be reduced by 30.3 and 28.2%, respectively. The latent heats and phase-change temperatures changed very little after 100 thermal cycles.

Solar Absorption

Solar energy is one of the best sources of renewable energy with minimal environmental impact. The conventional direct absorption solar collector is a well-established technology, and it has been proposed for a variety of applications such as water heating; however, the efficiency of these collectors is limited by the absorption properties of the working fluid, which is very poor for typical fluids used in solar collectors. Recently, this technology has been combined with the emerging technologies of nanofluids and liquid-nanoparticle suspensions to create a new class of nanofluid-based solar collectors. Otanicar et al. reported the experimental results on solar collectors based on nanofluids made from a variety of nanoparticles (CNTs, graphite, and silver) [81]. The efficiency improvement was up to 5% in solar thermal collectors by utilizing nanofluids as the absorption media. In addition, they compared the experimental data with a numerical model of a solar collector with direct absorption nanofluids. The experimental and numerical results demonstrated an initial rapid increase in efficiency with volume fraction, followed by a leveling off in efficiency as volume fraction continues to increase. Theoretical investigation on the feasibility of using a nonconcentrating direct absorption solar collector showed that the presence of nanoparticles increased the absorption of incident radiation by more than nine times over that of pure water [82]. Under the similar operating conditions, the efficiency of an absorption solar collector using nanofluid as the working fluid was found to be up to 10% higher (on an absolute basis) than that of a flat-plate collector. Otanicar and Golden evaluated the overall economic and environmental impacts of the technology in contrast with conventional solar collectors using the life-cycle assessment methodology [83]. Results showed that for the current cost of nanoparticles the nanofluid-based solar collector had a slightly longer payback period but at the end of its useful life has the same economic saving as a conventional solar collector. Sani et al. investigated the optical and thermal properties of nanofluids consisting of aqueous suspensions of single-wall carbon nanohorns [84]. The observed nanoparticle-induced differences in optical properties appeared promising, leading to a considerably higher sunlight absorption. Both these effects, together with the possible chemical functionalization of carbon nanohorns, make this new kind of nanofluids very interesting for increasing the overall efficiency of the sunlight exploiting device.

Mechanical Applications

Why nanofluids have great friction reduction properties? Nanoparticles in nanofluids form a protective film with low hardness and elastic modulus on the worn surface can be considered as the main reason that some nanofluids exhibit excellent lubricating properties.

Magnetic fluids are kinds of special nanofluids. Magnetic liquid rotary seals operate with no maintenance and extremely low leakage in a very wide range of applications, and it utilizing the property magnetic properties of the magnetic nanoparticles in liquid.

Friction Reduction

Advanced lubricants can improve productivity through energy saving and reliability of engineered systems. Tribological research heavily emphasizes reducing friction and wear. Nanoparticles have attracted much interest in recent years due to their excellent load-carrying capacity, good extreme pressure and friction reducing properties. Zhou et al. evaluated the tribological behavior of Cu nanoparticles in oil on a four-ball machine. The results showed that Cu nanoparticles as an oil additive had better friction-reduction and antiwear properties than zinc dithiophosphate, especially at high applied load. Meanwhile, the nanoparticles could also strikingly improve the load-carrying capacity of the base oil [85]. Dispersion of solid particles was found to play an important role, especially when a slurry layer was formed. Water-based Al_2O_3 and diamond nanofluids were applied in the minimum quantity lubrication (MQL) grinding process of cast iron. During the nanofluid MQL grinding, a dense and hard slurry layer was formed on the wheel surface and could benefit the grinding performance. Nanofluids showed the benefits of reducing grinding forces, improving surface roughness, and preventing workpiece burning. Compared to dry grinding, MQL grinding could significantly reduce the grinding temperature [86]. Wear and friction properties of surface modified Cu nanoparticles, as 50CC oil additive were studied. The higher the oil temperature applied, the better the tribological properties of Cu nanoparticles were. It could be inferred that a thin copper protective film with lower elastic modulus and hardness was formed on the worn surface, which resulted in the good tribological performances of Cu nanoparticles, especially when the oil temperature was higher [87]. Yu et al. firstly reported that room temperature ionic liquid multiwalled carbon nanotubes composite was evaluated as lubricant additive in ionic liquid due to their excellent dispersibility and that the composite showed good friction-reduction and antiwear properties in friction process [88]. Wang et al. studied the tribological properties of ionic liquid-based nanofluids containing functionalized MWNTs under loads in the range of 200–800 N [89], indicating that the nanofluids exhibited preferable friction-reduction properties under 800 N and remarkable antiwear properties with use of reasonable concentrations. Magnetic nanoparticle $Mn_{0.78}Zn_{0.22}Fe_2O_4$ was also an efficient lubricant additive. When used as a lubricant additive in 46 turbine oil, it could improve the wear resistance, load-carrying capacity, and antifriction ability of base oil, and the decreasing percentage of wear scar diameter was 25.45% compared to the base oil. This was a typical self-repair phenomenon [90]. Chen et al. reported on dispersion stability enhancement and self-repair principle discussion of ultrafine-tungsten disulfide in green lubricating oil [91]. Ultrafine-tungsten disulfide particulates could fill and level up the furrows on abrasive surfaces, repairing abrasive surface well. What is more, ultrafine-tungsten disulfide particulates could form a WS_2 film with low shear stress by adsorbing and depositing in the hollowness of abrasive surface, making the abrasive surface be more smooth, and the FeS film formed in tribochemical reaction could protect the abrasive surface further, all of which realize the self-repair to abrasive surface. The tribological properties of liquid paraffin with SiO_2 nanoparticles additive made by a sol-gel method was investigated by Peng et al. [92]. The optimal concentrations of SiO_2 nanoparticles in liquid paraffin was associated with better tribological properties than pure paraffin oil, and an antiwear ability that depended on the particle size, and oleic acid surface-modified SiO_2 nanoparticles with an average diameter of 58 nm provided better tribological properties in load-carrying capacity, antiwear and friction-reduction than pure liquid paraffin. Nanoparticles can easily penetrate into the rubbing surfaces because of their nanoscale. During the frictional process, the thin physical tribofilm of the nanoparticles forms between rubbing surfaces, which cannot only bear the load, but also separates the rubbing surfaces. The spherical SiO_2 nanoparticles could roll between the rubbing surfaces in sliding friction, and the originally pure sliding friction becomes mixed sliding and rolling friction. Therefore, the friction coefficient declines markedly and then remains constant.

Magnetic Sealing

Magnetic fluids (ferromagnetic fluid) are kinds of special nanofluids. They are stable colloidal suspensions of small magnetic particles such as magnetite (Fe_3O_4). The properties of the magnetic nanoparticles, the magnetic component of magnetic nanofluids, may be tailored by varying their size and adapting their surface coating in order to meet the requirements of colloidal stability of magnetic nanofluids with nonpolar and polar carrier liquids [93]. Comparing with the mechanical sealing, magnetic sealing offers a cost-effective solution to environmental and hazardous-gas sealing in a wide variety of industrial rotation equipment with high-speed capability, low-friction power losses, and long life and high reliability [94]. A ring magnet forms part of a magnetic circuit in which an intense magnetic field is established in the gaps between the teeth on a magnetically permeable shaft and the surface of an opposing pole block. Ferrofluid introduced into the gaps forms discrete liquid rings capable of supporting a pressure difference while maintaining zero leakage. The seals operate without wear as the shaft rotates, because the mechanical moving parts do not touch.

With these unique characteristics, sealing liquids with magnetic fluids can be applied in many application areas. It is reported that an iron particle dispersed magnetic fluids was utilized in the sealing of a high-rotation pump. The sealing holds pressure of 618 kPa with an 1800 r/min [95]. Mitamura et al. studied the application of a magnetic fluid seal to rotary blood pumps. The developed magnetic fluid seal worked for over 286 days in a continuous flow condition, for 24 days (on-going) in a pulsatile flow condition and for 24 h (electively terminated) in blood flow [96]. Ferrocobalt magnetic fluid was used for oil sealing, and the holding pressure is 25 times as high as that of a conventional magnetite sealing [97].

Biomedical Application

For some special kinds of nanoparticles, they have antibacterial activities or drug-delivery properties, so the nanofluids containing these nanoparticles will exhibit some relevant properties.

Antibacterial Activity

Organic antibacterial materials are often less stable particularly at high temperatures or pressures. As a consequence, inorganic materials such as metal and metal oxides have attracted lots of attention over the past decade due to their ability to withstand harsh process conditions. The antibacterial behavior of ZnO nanofluids shows that the ZnO nanofluids have bacteriostatic activity against [98]. Electrochemical measurements suggest some direct interaction between ZnO nanoparticles and the bacteria membrane at high ZnO concentrations. Jalal et al. prepared ZnO nanoparticles via a green method. The antibacterial activity of suspensions of ZnO nanoparticles against Escherichia coli (E. coli) has been evaluated by estimating the reduction ratio of the bacteria treated with ZnO. Survival ratio of bacteria decreases with increasing the concentrations of ZnO nanofluids and time [99]. Further investigations have clearly demonstrated that ZnO nanoparticles have a wide range of antibacterial effects on a number of other microorganisms. The antibacterial activity of ZnO may be dependent on the size and the presence of normal visible light [100]. Recent research showed that ZnO nanoparticles exhibited impressive antibacterial properties against an important foodborne pathogen, E. coli O157:H7, and the inhibitory effects increased as the concentrations of ZnO nanoparticles increased. ZnO nanoparticles changed the cell membrane components including lipids and proteins. ZnO nanoparticles could distort bacterial cell membrane, leading to loss of intracellular components, and ultimately the death of cells, considered as an effective antibacterial agent for protecting agricultural and food safety [101].

The antibacterial activity research of CuO nanoparticles showed that they possessed antibacterial activity against four bacterial strains. The size of nanoparticles was less than that of the pore size in the bacteria, and thus, they had a unique property of crossing the cell membrane without any hindrance. It could be hypothesized that these nanoparticles formed stable complexes with vital enzymes inside cells which hampered cellular functioning resulting in their death [102]. Bulk equivalents of these products showed no inhibitory activity, indicating that particle size was determinant in activity [103]. Lee et al. reported the antibacterial efficacy of nanosized silver colloidal solution on the cellulosic and synthetic fabrics [104]. The antibacterial treatment of the textile fabrics was easily achieved by padding them with nanosized silver colloidal solution. The antibacterial efficacy of the fabrics was maintained after many times laundering. Silver colloid is an efficient antibacterial agent. The silver colloid prepared by a one-step synthesis showed high antimicrobial and bactericidal activity against Gram-positive and Gram-negative bacteria, including highly multiresistant strains such as methicillin-resistant staphylococcus aureus. The antibacterial activity of silver nanoparticles was found to be dependent on the size of silver particles. A very low concentration of silver gave antibacterial performance [105]. The aqueous suspensions of fullerenes and nano-TiO_2 can produce reactive oxygen species (ROS). Bacterial (E. coli) toxicity tests suggested that unlike nano-TiO_2 which was exclusively phototoxic, the antibacterial activity of fullerene suspensions was linked to ROS production. Nano-TiO_2 may be more efficient for water treatment involving UV or solar energy, to enhance contaminant oxidation and perhaps for disinfection. However, fullerol and PVP/C_{60} may be useful as water treatment agents targeting specific pollutants or microorganisms that are more sensitive to either superoxide or singlet oxygen [106]. Lyon and Alvarez proposed that C_{60} suspensions exerted ROS-independent oxidative stress in bacteria, with evidence of protein oxidation, changes in cell membrane potential, and interruption of cellular respiration. This mechanism requires direct contact between the nanoparticle and the bacterial cell and differs from previously reported nanomaterial antibacterial mechanisms that involve ROS generation (metal oxides) or leaching of toxic elements (nanosilver) [107].

Nanodrug Delivery

Over the last few decades, colloidal drug delivery systems have been developed in order to improve the efficiency and the specificity of drug action [108]. The small-size, customized surface improved solubility, and multifunctionality of nanoparticles opens many doors and creates new biomedical applications. The novel properties of nanoparticles offer the ability to interact with complex cellular functions in new ways [109]. Gold nanoparticles provide nontoxic carriers for drug- and gene-delivery applications. With these systems, the gold core imparts stability to the assembly, while the monolayer allows tuning of surface properties such as charge and hydrophobicity. Another attractive feature of gold nanoparticles is their interaction with thiols, providing an effective and selective means of controlled intracellular release [110]. Nakano et al. proposed the drug-delivery system using nanomagnetic fluid [111], which targeted and concentrated drugs using a ferrofluid cluster composed of magnetic nanoparticles. The potential of magnetic nanoparticles stems from the intrinsic properties of their magnetic cores combined with their drug-loading capability and the biochemical properties that can be bestowed on them by means of a suitable coating. CNT has emerged as a new alternative and efficient tool for transporting and Trans locating therapeutic molecules. CNT can be functionalized with bioactive peptides, proteins, nucleic acids, and drugs and used to deliver their cargos to cells and organs. Because functionalised CNT display low toxicity and are not immunogenic, such systems hold great potential in the field of nanobiotechnology and nanomedicine [112, 113]. Pastorin et al. have developed a novel strategy for the functionalisation of CNTs with two different molecules using the 1, 3-dipolar cycloaddition of azomethine ylides [114]. The attachment of molecules that will target specific receptors on tumour cells will help improve the response to anticancer agents. Liu et al. have found that prefunctionalized CNTs can adsorb widely used aromatic molecules by simple mixing, forming "forest-scrub"-like assemblies on CNTs with PEG extending into water to impart solubility and aromatic molecules densely populating CNT sidewalls. The work establishes a novel, easy-to-make formulation of a SWNT-doxorubicin complex with extremely high drug loading efficiency [115].

In recent years, graphene based drug delivery systems have attracted more and more attention. In 2008, Sun et al. firstly reported the application of nanographene oxide (NGO) for cellular imaging and drug delivery [116]. They have developed functionalization chemistry in order to impart solubility and compatibility of NGO in biological environments. Simple physicosorption via ϖ-stacking can be used for loading doxorubicin, a widely used cancer drug onto NGO functionalized with antibody for selective killing of cancer cells in vitro. Functional nanoscale graphene oxide is found to be a novel nanocarrier for the loading and targeted delivery of anticancer drugs [117]. Controlled loading of two anticancer drugs onto the folic acid-conjugated NGO via ϖ-ϖ stacking and hydrophobic interactions demonstrated that NGO loaded with the two anticancer drugs showed specific targeting to MCF-7 cells (human breast cancer cells with folic acid receptors), and remarkably high cytotoxicity compared to NGO loaded with either doxorubicin or camptothecin only. The PEGylated (PEG: polyethylene glycol) nanographene oxide could be used for the delivery of water-insoluble cancer drugs [118]. PEGylated NGO readily complexes with a water-insoluble aromatic molecule SN38, a camptothecin analogue, via noncovalent van der Waals interaction. The NGO-PEG-SN38 complex exhibits excellent aqueous solubility and retains the high potency of free SN38 dissolved in organic solvents. Yang et al. found $GO-Fe_3O_4$ hybrid could be loaded with anticancer drug doxorubicin hydrochloride with a high loading capacity [119]. This $GO-Fe_3O_4$ hybrid showed superparamagnetic property and could congregate under acidic conditions and be redispersed reversibly under basic conditions. This pH-triggered controlled magnetic behavior makes this material a promising candidate for controlled targeted drug delivery.

Other Applications

Intensify Microreactors

The discovery of high enhancement of heat transfer in nanofluids can be applicable to the area of process intensification of chemical reactors through integration of the functionalities of reaction and heat transfer in compact multifunctional reactors. Fan et al. studied a nanofluid based on benign TiO_2 material dispersed in ethylene glycol in an integrated reactor-heat exchanger [120]. The overall heat transfer coefficient increase was up to 35% in the steady state continuous experiments. This resulted in a closer temperature control in the reaction of selective reduction of an aromatic aldehyde by molecular hydrogen and very rapid change in the temperature of reaction under dynamic reaction control.

Nanofluids as Vehicular Brake Fluids

A vehicle's kinetic energy is dispersed through the heat produced during the process of braking and this is transmitted throughout the brake fluid in the hydraulic braking system [39], and now, there is a higher demand for the properties of brake oils. Copper-oxide and aluminum-oxide based brake nanofluids were manufactured using the arc-submerged nanoparticle synthesis system and the plasma charging arc system, respectively [121, 122]. The two kinds of nanofluids both have enhanced properties such as a higher boiling point, higher viscosity, and a higher conductivity than that of traditional brake fluid. By yielding a higher boiling point, conductivity, and viscosity, the nanofluid brake oil will reduce the occurrence of vapor-lock and offer increased safety while driving.

Nanofluids-Based Microbial Fuel Cell

Microbial fuel cells (MFC) that utilize the energy found in carbohydrates, proteins, and other energy-rich natural products to generate electrical power have a promising future. The excellent performance of MFC depends on electrodes and electron mediator. Sharma ET al.constructed a novel microbial fuel cell (MFC) using novel electron mediators and CNT-based electrodes [123]. The novel mediators are nanofluids which were prepared by dispersing nanocrystalline platinum anchored CNTs in water. They compared the performance of the new E. coli-based MFC to the previously reported E. coli-based microbial fuel cells with neutral red and methylene blue electron mediators. The performance of the MFC using CNT-based nanofluids and CNT-based electrodes has been compared against plain graphite electrode-based MFC. CNT-based electrodes showed as high as ~6-fold increase in the power density compared to graphite electrodes. The work demonstrates the potential of noble metal nanoparticles dispersed on CNT-based MFC for the generation of high energies from even simple bacteria like E. coli.

Nanofluids with Unique Optical Properties

Optical filters are used to select different wavelengths of light. The ferrofluid-based optical filter has tunable properties. The desired central wavelength region can be tuned by an external magnetic field. Philip et al. developed a ferrofluid-based emulsion for selecting different bands of wavelengths in the UV, visible, and IR regions [124]. The desired range of wavelengths, bandwidth, and percentage of reflectivity could be easily controlled by using suitably tailored ferrofluid emulsions. Mishra et al. developed nanofluids with selective visible colors in gold nanoparticles embedded in polymer molecules of polyvinyl pyrrolidone (PVP) in water [125]. They compared the developments in the apparent visible colors in forming the Au-PVP nanofluids of 0.05, 0.10, 0.50, and 1.00 wt% Au contents. The surface plasmon bands, which occurs over 480–700 nm, varies sensitively in its position as well as the intensity when varying the Au content 0-1 wt%.

CONCLUSIONS AND FUTURE WORK

Many interesting properties of nanofluids have been reported in the past decades. This paper presents an overview of the recent developments in the study of nanofluids, including the preparation methods, the evaluation methods for their stability, the ways to enhance their stability, the stability mechanisms, and their potential applications in heat transfer intensification, mass transfer enhancement, energy fields, mechanical fields, biomedical fields, and so forth.

Although nanofluids have displayed enormously exciting potential applications, some vital hinders also exist before commercialization of nanofluids. The following key issues should receive greater attention in the future. Firstly, further experimental and theoretical research is required to find the major factors influencing the performance of nanofluids. Up to now, there is a lack of agreement between experimental results from different groups, so it is important to systematically identify these factors. The detailed and accurate structure characterizations of the suspensions may be the key to explain the discrepancy in the experimental data. Secondly, increase in viscosity by the use of nanofluids is an important drawback due to the associated increase in pumping power. The applications for nanofluids with low viscosity and high conductivity are promising. Enhancing the compatibility between nanomaterials and the base fluids through modifying the interface properties of two phases may be one of the solution routes. Thirdly, the shape of the additives in nanofluids is very important for the properties; therefore, the new nanofluid synthesis approaches with controllable microscope structure will be an interesting research work. Fourthly, stability of the suspension is a crucial issue for both scientific research and practical applications. The stability of nanofluids, especially the long-term stability, the stability in the practical conditions, and the stability after thousands of thermal cycles should be paid more attention. Fifthly, there is a lack of investigation of the thermal performance of nanofluids at high temperatures, which may widen the possible application areas of nanofluids, like in high-temperature solar energy absorption and high-temperature energy storage. At the same time, high temperature may accelerate the degradation of the surfactants used as dispersants in nanofluids and may produce more foams. These factors should be taken into account. Finally, the properties of nanofluids strongly depend on the shape and property of the additive. Xie's findings indicated that thermal conductivity enhancement was adjusted by ball milling and cutting the treated CNTs suspended in the nanofluids to relatively straight CNTs with an appropriate length distribution. They proposed the concept of straightness ratio to explain the facts (Figure 3). Nanofluid research can be enrichened and extended through exploring new nanomaterials. For example, the newly discovered 2D monatomic sheet graphene is a promising candidate material to enhance the thermal conductivity of the base fluid [126, 127], as shown in Figure 4. The concept of nanofluids is extended by the use of phase change materials, which goes well beyond simply increasing the thermal conductivity of a fluid [128]. It is found that the indium/polyalphaolefin phase change nanofluid exhibits simultaneously enhanced thermal conductivity and specific heat.

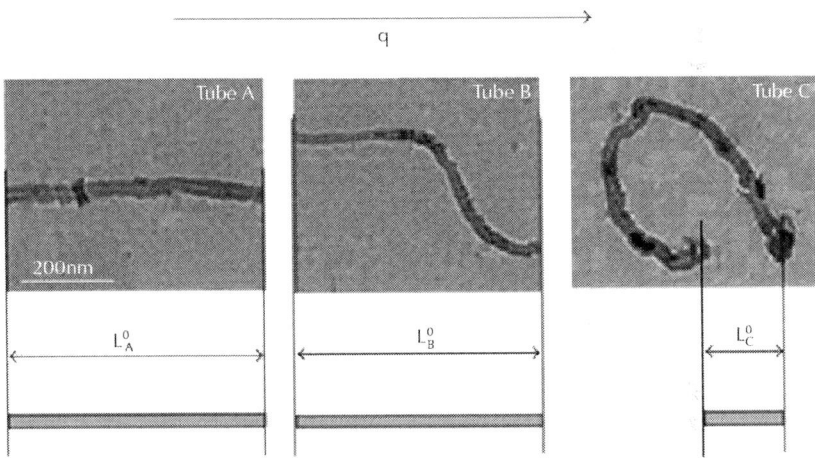

Figure 3: Actual nonstraight CNTs (left two) and equivalent straight thermal passages (right).

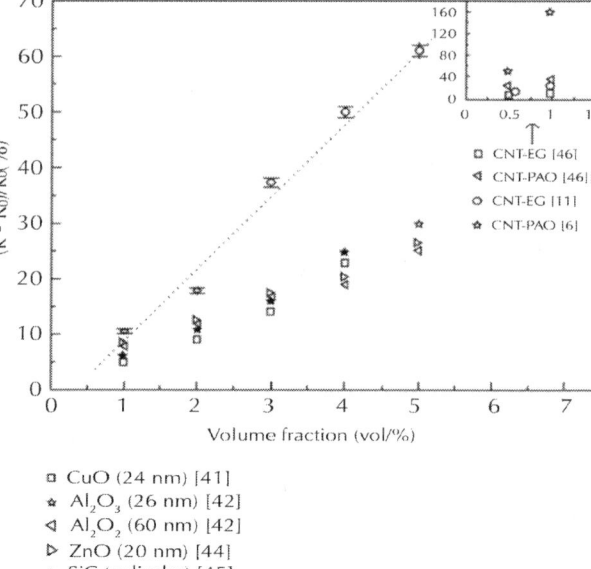

CuO (24 nm) [41]
Al$_2$O$_3$ (26 nm) [42]
Al$_2$O$_2$ (60 nm) [42]
ZnO (20 nm) [44]
SiC (cylinder) [45]
GON-EG

Figure 4: Thermal conductivity enhancement ratios of EG-based nanofluids as a function of loading. The inset shows the thermal conductivity enhancement ratios of nanofluids containing CNTs.

ACKNOWLEDGMENTS

The work was supported by New Century Excellent Talents in University (NECT-10-883), the Program for Professor of Special Appointment (Eastern Scholar) at Shanghai Institutions of Higher Learning, and partly by National Natural Science Foundation of China (51106093).

REFERENCES

1.   V. Trisaksri and S. Wongwises, "Critical review of heat transfer characteristics of nanofluids," Renewable and Sustainable Energy Reviews, vol. 11, no. 3, pp. 512–523, 2007. · ·

2.   S. Özerinç, S. Kakaç, and A. G. Yazıcıoĝlu, "Enhanced thermal conductivity of nanofluids: a state-of-the-art review," Microfluidics and Nanofluidics, vol. 8, no. 2, pp. 145–170, 2010. ·

3.   X. Q. Wang and A. S. Mujumdar, "Heat transfer characteristics of nanofluids: a review,"International Journal of Thermal Sciences, vol. 46, no. 1, pp. 1–19, 2007. ·

4.   X. Q. Wang and A. S. Mujumdar, "A review on nanofluids—part I: theoretical and numerical investigations," Brazilian Journal of Chemical Engineering, vol. 25, no. 4, pp. 613–630, 2008.

5.   Y. Li, J. Zhou, S. Tung, E. Schneider, and S. Xi, "A review on development of nanofluid preparation and characterization," Powder Technology, vol. 196, no. 2, pp. 89–101, 2009. · ·

6.   S. Kakaç and A. Pramuanjaroenkij, "Review of convective heat transfer enhancement with nanofluids," International Journal of Heat and Mass Transfer, vol. 52, no. 13-14, pp. 3187–3196, 2009. · ·

7.   J. A. Eastman, S. U. S. Choi, S. Li, W. Yu, and L. J. Thompson, "Anomalously increased effective thermal conductivities of ethylene glycol-based nanofluids containing copper nanoparticles," Applied Physics Letters, vol. 78, no. 6, pp. 718–720, 2001. · ·

8.   C. H. Lo, T. T. Tsung, and L. C. Chen, "Shape-controlled synthesis of Cu-based nanofluid using submerged arc nanoparticle synthesis system (SANSS)," Journal of Crystal Growth, vol. 277, no. 1–4, pp. 636–642, 2005. · ·

9.   C. H. Lo, T. T. Tsung, L. C. Chen, C. H. Su, and H. M. Lin, "Fabrication of copper oxide nanofluid using submerged arc nanoparticle synthesis system (SANSS)," Journal of Nanoparticle Research, vol. 7, no. 2-3, pp. 313–320, 2005. · ·

10. H. T. Zhu, Y. S. Lin, and Y. S. Yin, "A novel one-step chemical method for preparation of copper nanofluids," Journal of Colloid and Interface Science, vol. 277, no. 1, pp. 100–103, 2004. · ·

11. H. Bönnemann, S. S. Botha, B. Bladergroen, and V. M. Linkov, "Monodisperse copper- and silver-nanocolloids suitable for heat-conductive fluids," Applied Organometallic Chemistry, vol. 19, no. 6, pp. 768–773, 2005. · ·

12. A. K. Singh and V. S. Raykar, "Microwave synthesis of silver nanofluids with polyvinylpyrrolidone (PVP) and their transport

properties," Colloid and Polymer Science, vol. 286, no. 14-15, pp. 1667–1673, 2008. · ·

13. A. Kumar, H. Joshi, R. Pasricha, A. B. Mandale, and M. Sastry, "Phase transfer of silver nanoparticles from aqueous to organic solutions using fatty amine molecules," Journal of Colloid and Interface Science, vol. 264, no. 2, pp. 396–401, 2003. · ·

14. W. Yu, H. Xie, X. Wang, and X. Wang, "Highly efficient method for preparing homogeneous and stable colloids containing graphene oxide," Nanoscale Research Letters, vol. 6, p. 47, 2011.

15. H. T. Zhu, C. Y. Zhang, Y. M. Tang, and J. X. Wang, "Novel synthesis and thermal conductivity of CuO nanofluid," Journal of Physical Chemistry C, vol. 111, no. 4, pp. 1646–1650, 2007. · ·

16. Y. Chen and X. Wang, "Novel phase-transfer preparation of monodisperse silver and gold nanoparticles at room temperature," Materials Letters, vol. 62, no. 15, pp. 2215–2218, 2008. ·

17. X. Feng, H. Ma, S. Huang et al., "Aqueous-organic phase-transfer of highly stable gold, silver, and platinum nanoparticles and new route for fabrication of gold nanofilms at the oil/water interface and on solid supports," Journal of Physical Chemistry B, vol. 110, no. 25, pp. 12311–12317, 2006. · ·

18. W. Yu, H. Xie, L. Chen, and Y. Li, "Enhancement of thermal conductivity of kerosene-based Fe3O4 nanofluids prepared via phase-transfer method," Colloids and Surfaces A, vol. 355, no. 1–3, pp. 109–113, 2010. · ·

19. L. Wang and J. Fan, "Nanofluids research: key issues," Nanoscale Research Letters, vol. 5, no. 8, pp. 1241–1252, 2010. · ·

20. X. Wei and L. Wang, "Synthesis and thermal conductivity of microfluidic copper nanofluids," Particuology, vol. 8, no. 3, pp. 262–271, 2010. · ·

21. X. Li, D. Zhu, and X. Wang, "Evaluation on dispersion behavior of the aqueous copper nano-suspensions," Journal of Colloid and Interface Science, vol. 310, no. 2, pp. 456–463, 2007. · ·

22. H. Zhu, C. Zhang, Y. Tang, J. Wang, B. Ren, and Y. Yin, "Preparation and thermal conductivity of suspensions of graphite nanoparticles," Carbon, vol. 45, no. 1, pp. 226–228, 2007. · ·

23. D. Li and R. B. Kaner, "Processable stabilizer-free polyaniline nanofiber aqueous colloids," Chemical Communications, vol. 14, no. 26, pp. 3286–3288, 2005. · ·

24. H. J. Kim, I. C. Bang, and J. Onoe, "Characteristic stability of bare Au-water nanofluids fabricated by pulsed laser ablation in liquids," Optics and Lasers in Engineering, vol. 47, no. 5, pp. 532–538, 2009. · ·

25. X. J. Wang, X. Li, and S. Yang, "Influence of pH and SDBS on the stability and thermal conductivity of nanofluids," Energy and Fuels, vol. 23, no. 5, pp. 2684–2689, 2009.

26. D. Zhu, X. Li, N. Wang, X. Wang, J. Gao, and H. Li, "Dispersion behavior and thermal conductivity characteristics of Al_2O_3-H_2O nanofluids," Current Applied Physics, vol. 9, no. 1, pp. 131–139, 2009. ·

27. L. Chen and H. Xie, "Properties of carbon nanotube nanofluids stabilized by cationic gemini surfactant," Thermochimica Acta, vol. 506, no. 1-2, pp. 62–66, 2010. ·

28. J. Huang, X. Wang, Q. Long, X. Wen, Y. Zhou, and L. Li, "Influence of pH on the stability characteristics of nanofluids," in Proceedings of the Symposium on Photonics and Optoelectronics (SOPO ‹09), 2009. ·

29. M. Farahmandjou, S. A. Sebt, S. S. Parhizgar, P. Aberomand, and M. Akhavan, "Stability investigation of colloidal FePt nanoparticle systems by spectrophotometer analysis,"Chinese Physics Letters, vol. 26, no. 2, Article ID 027501, 2009. · ·

30. Y. Hwang, J. K. Lee, C. H. Lee et al., "Stability and thermal conductivity characteristics of nanofluids," Thermochimica Acta, vol. 455, no. 1-2, pp. 70–74, 2007. · ·

31. L. Chen, H. Xie, Y. Li, and W. Yu, "Nanofluids containing carbon nanotubes treated by mechanochemical reaction," Thermochimica Acta, vol. 477, no. 1-2, pp. 21–24, 2008. ·

32. X. Yang and Z. H. Liu, "A kind of nanofluid consisting of surface-functionalized nanoparticles," Nanoscale Research Letters, vol. 5, no. 8, pp. 1324–1328, 2010.

33. L. Chen and H. Xie, "Surfactant-free nanofluids containing double- and single-walled carbon nanotubes functionalized by a

wet-mechanochemical reaction," Thermochimica Acta, vol. 497, no. 1-2, pp. 67–71, 2010. ·

34. K. A. Wepasnick, B. A. Smith, J. L. Bitter, and D. H. Fairbrother, "Chemical and structural characterization of carbon nanotube surfaces," Analytical and Bioanalytical Chemistry, vol. 396, no. 3, pp. 1003–1014, 2010. · ·

35. Q. Yu, Y. J. Kim, and H. Ma, "Nanofluids with plasma treated diamond nanoparticles,"Applied Physics Letters, vol. 92, no. 10, Article ID 103111, 2008. ·

36. I. M. Joni, A. Purwanto, F. Iskandar, and K. Okuyama, "Dispersion stability enhancement of titania nanoparticles in organic solvent using a bead mill process,"Industrial and Engineering Chemistry Research, vol. 48, no. 15, pp. 6916–6922, 2009. · ·

37. E. Tang, G. Cheng, X. Ma, X. Pang, and Q. Zhao, "Surface modification of zinc oxide nanoparticle by PMAA and its dispersion in aqueous system," Applied Surface Science, vol. 252, no. 14, pp. 5227–5232, 2006. · ·

38. T. Missana and A. Adell, "On the applicability of DLVO theory to the prediction of clay colloids stability," Journal of Colloid and Interface Science, vol. 230, no. 1, pp. 150–156, 2000. · ·

39. I. Popa, G. Gillies, G. Papastavrou, and M. Borkovec, "Attractive and repulsive electrostatic forces between positively charged latex particles in the presence of anionic linear polyelectrolytes," Journal of Physical Chemistry B, vol. 114, no. 9, pp. 3170–3177, 2010. · ·

40. H. Kamiya, Y. Fukuda, Y. Suzuki, M. Tsukada, T. Kakui, and M. Naito, "Effect of polymer dispersant structure on electrosteric interaction and dense alumina suspension behavior," Journal of the American Ceramic Society, vol. 82, no. 12, pp. 3407–3412, 1999.

41. K. V. Wong and O. de Leon, "Applications of nanofluids: current and future," Advances in Mechanical Engineering, vol. 2010, Article ID 519659, 11 pages, 2010.

42. G. Donzelli, R. Cerbino, and A. Vailati, "Bistable heat transfer in a nanofluid," Physical Review Letters, vol. 102, no. 10, Article ID 104503, 2009. · ·

43. M. Arruebo, R. Fernández-Pacheco, M. R. Ibarra, and J. Santamaría, "Magnetic nanoparticles for drug delivery," Nano Today, vol. 2, no. 3, pp. 22–32, 2007. ·

44. W. Yu, D. M. France, D. Singh, E. V. Timofeeva, D. S. Smith, and J. L. Routbort, "Mechanisms and models of effective thermal conductivities of nanofluids," Journal of Nanoscience and Nanotechnology, vol. 10, no. 8, pp. 4824–4849, 2010. ·

45. K. Q. Ma and J. Liu, "Nano liquid-metal fluid as ultimate coolant," Physics Letters Section a, vol. 361, no. 3, pp. 252–256, 2007. · ·

46. G. Paul, M. Chopkar, I. Manna, and P. K. Das, "Techniques for measuring the thermal conductivity of nanofluids: a review," Renewable and Sustainable Energy Reviews, vol. 14, p. 1913, 2010.

47. H. Xie, W. Yu, and W. Chen, "MgO nanofluids: higher thermal conductivity and lower viscosity among ethylene glycol-based nanofluids containing oxide nanoparticles,"Journal of Experimental Nanoscience, vol. 5, no. 5, pp. 463–472, 2010. · ·

48. S. P. Jang and S. U. S. Choi, "Cooling performance of a microchannel heat sink with nanofluids," Applied Thermal Engineering, vol. 26, no. 17-18, pp. 2457–2463, 2006.

49. C. T. Nguyen, G. Roy, N. Galanis, and S. Suiro, "Heat transfer enhancement by using Al_2O_3-water nanofluid in a liquid cooling system for microprocessors," in Proceedings of the 4th WSEAS International Conference on Heat Transfer, Thermal Engineering and Environment, pp. 103–108, Elounda, Greece, August 2006.

50. H. Shokouhmand, M. Ghazvini, and J. Shabanian, "Performance analysis of using nanofluids in microchannel heat sink in different flow regimes and its simulation using artificial neural network," in Proceedings of the World Congress on Engineering (WCE '08), vol. 3, London, UK, July 2008.

51. C. Y. Tsaia, H. T. Chiena, P. P. Dingb, B. Chanc, T. Y. Luhd, and P. H. Chena, "Effect of structural character of gold nanoparticles in nanofluid on heat pipe thermal performance," Materials Letters, vol. 58, p. 1461, 2004.

52. Y. T. Chen, W. C. Wei, S. W. Kang, and C. S. Yu, "Effect of nanofluid on flat heat pipe thermal performance," in Proceedings of the 24th IEEE Semiconductor Thermal Measurement and Management Symposium (SEMI-THERM '08), March 2006.

53.   H. B. Ma, C. Wilson, B. Borgmeyer et al., "Effect of nanofluid on the heat transport capability in an oscillating heat pipe," Applied Physics Letters, vol. 88, no. 14, Article ID 143116, 2006. · ·

54.   S. W. Kang, W. C. Wei, S. H. Tsai, and C. C. Huang, "Experimental investigation of nanofluids on sintered heat pipe thermal performance," Applied Thermal Engineering, vol. 29, no. 5-6, pp. 973–979, 2009. ·

55.   P. Naphon, P. Assadamongkol, and T. Borirak, "Experimental investigation of titanium nanofluids on the heat pipe thermal efficiency," International Communications in Heat and Mass Transfer, vol. 35, no. 10, pp. 1316–1319, 2008.

56.   H. Xie and L. Chen, "Adjustable thermal conductivity in carbon nanotube nanofluids,"Physics Letters Section a, vol. 373, no. 21, pp. 1861–1864, 2009. · ·

57.   H. Xie, W. Yu, and Y. Li, "Thermal performance enhancement in nanofluids containing diamond nanoparticles," Journal of Physics D, vol. 42, no. 9, Article ID 095413, 2009. ·

58.   W. Yu, H. Xie, L. Chen, and Y. Li, "Investigation of thermal conductivity and viscosity of ethylene glycol based ZnO nanofluid," Thermochimica Acta, vol. 491, no. 1-2, pp. 92–96, 2009. ·

59.   W. Yu, D. M. France, S. U. S. Choi, and J. L. Routbort, "Review and assessment of nanofluid technology for transportation and other applications," Tech. Rep. 78, ANL/ESD/07-9, Argonne National Laboratory, 2007.

60.   M. Kole and T. K. Dey, "Thermal conductivity and viscosity of Al2O3 nanofluid based on car engine coolant," Journal of Physics D, vol. 43, no. 31, Article ID 315501, 2010. ·

61.   S. C. Tzeng, C. W. Lin, and K. D. Huang, "Heat transfer enhancement of nanofluids in rotary blade coupling of four-wheel-drive vehicles," Acta Mechanica, vol. 179, no. 1-2, pp. 11–23, 2005. · ·

62.   D. Singh, J. Toutbort, G. Chen, et al., "Heavy vehicle systems optimization merit review and peer evaluation," Annual Report, Argonne National Laboratory, 2006.

63.   http://www.labnews.co.uk/feature_archive.php/5449/5/keeping-it-cool.

64. J. Routbort, et al., Argonne National Lab, Michellin North America, St. Gobain Corp., 2009, http://www1.eere.energy.gov/industry/nanomanufacturing/pdfs/nanofluids industrial cooling. pdf.

65. http://96.30.12.13/execsumm/VU0319–Nanofluid%20for%20 Cooling%20Enhancement%20of%20Electrical%20Power%20 Equipment.pdf.

66. I. C. Nelson, D. Banerjee, and R. Ponnappan, "Flow loop experiments using polyalphaolefin nanofluids," Journal of Thermophysics and Heat Transfer, vol. 23, no. 4, pp. 752–761, 2009. · ·

67. D. P. Kulkarni, D. K. Das, and R. S. Vajjha, "Application of nanofluids in heating buildings and reducing pollution," Applied Energy, vol. 86, no. 12, pp. 2566–2573, 2009. · ·

68. J. Boungiorno, L. W. Hu, S. J. Kim, R. Hannink, B. Truong, and E. Forrest, "Nanofluids for enhanced economics and safety of nuclear reactors: an evaluation of the potential features issues, and research gaps," Nuclear Technology, vol. 162, no. 1, pp. 80–91, 2008.

69. S. M. You, J. H. Kim, and K. H. Kim, "Effect of nanoparticles on critical heat flux of water in pool boiling heat transfer," Applied Physics Letters, vol. 83, no. 16, pp. 3374–3376, 2003. ·

70. P. Vassallo, R. Kumar, and S. D›Amico, "Pool boiling heat transfer experiments in silica-water nano-fluids," International Journal of Heat and Mass Transfer, vol. 47, no. 2, pp. 407–411, 2004. · ·

71. J. K. Kim, J. Y. Jung, and Y. T. Kang, "The effect of nano-particles on the bubble absorption performance in a binary nanofluid," International Journal of Refrigeration, vol. 29, no. 1, pp. 22–29, 2006. ·

72. J. K. Kim, J. Y. Jung, and Y. T. Kang, "Absorption performance enhancement by nano-particles and chemical surfactants in binary nanofluids," International Journal of Refrigeration, vol. 30, no. 1, pp. 50–57, 2007. ·

73. J. Kim, Y. T. Kang, and C. K. Choi, "Soret and Dufour effects on convective instabilities in binary nanofluids for absorption application," International Journal of Refrigeration, vol. 30, no. 2, pp. 323–328, 2007. ·

74. X. Ma, F. Su, J. Chen, and Y. Zhang, "Heat and mass transfer enhancement of the bubble absorption for a binary nanofluid," Journal of Mechanical Science and Technology, vol. 21, p. 1813, 2007.

75. X. Ma, F. Su, J. Chen, T. Bai, and Z. Han, "Enhancement of bubble absorption process using a CNTs-ammonia binary nanofluid," International Communications in Heat and Mass Transfer, vol. 36, no. 7, pp. 657–660, 2009. ·

76. S. Komati and A. K. Suresh, "CO2 absorption into amine solutions: a novel strategy for intensification based on the addition of ferrofluids," Journal of Chemical Technology and Biotechnology, vol. 83, no. 8, pp. 1094–1100, 2008. · ·

77. L. Yang, K. Du, B. Cheng, and Y. Jiang, "The influence of Al2O3 nanofluid on the falling film absorption with ammonia-water," in Proceedings of the Asia-Pacific Power and Energy Engineering Conference (APPEEC ‹10), 2010. ·

78. M. F. Demirbas, "Thermal energy storage and phase change materials: an overview,"Energy Sources Part B, vol. 1, no. 1, pp. 85–95, 2006. · ·

79. S. Wu, D. Zhu, X. Zhang, and J. Huang, "Preparation and melting/freezing characteristics of Cu/paraffin nanofluid as phase-change material (PCM)," Energy and Fuels, vol. 24, no. 3, pp. 1894–1898, 2010. · ·

80. Y. D. Liu, Y. G. Zhou, M. W. Tong, and X. S. Zhou, "Experimental study of thermal conductivity and phase change performance of nanofluids PCMs," Microfluidics and Nanofluidics, vol. 7, no. 4, pp. 579–584, 2009. · ·

81. T. P. Otanicar, P. E. Phelan, R. S. Prasher, G. Rosengarten, and R. A. Taylor, "Nanofluid-based direct absorption solar collector," Journal of Renewable and Sustainable Energy, vol. 2, no. 3, Article ID 033102, 13 pages, 2010. ·

82. H. Tyagi, P. Phelan, and R. Prasher, "Predicted efficiency of a low-temperature Nanofluid-based direct absorption solar collector," Journal of Solar Energy Engineering, vol. 131, no. 4, pp. 0410041–0410047, 2009. · ·

83. T. P. Otanicar and J. S. Golden, "Comparative environmental and economic analysis of conventional and nanofluid solar hot water

technologies," Environmental Science and Technology, vol. 43, no. 15, pp. 6082–6087, 2009. · ·

84. E. Sani, S. Barison, C. Pagura et al., "Carbon nanohorns-based nanofluids as direct sunlight absorbers," Optics Express, vol. 18, p. 4613, 2010.

85. J. Zhou, Z. Wu, Z. Zhang, W. Liu, and Q. Xue, "Tribological behavior and lubricating mechanism of Cu nanoparticles in oil," Tribology Letters, vol. 8, no. 4, pp. 213–218, 2000.

86. B. Shen, A. J. Shih, and S. C. Tung, "Application of nanofluids in minimum quantity lubrication grinding," Tribology Transactions, vol. 51, no. 6, pp. 730–737, 2008.

87. H. L. Yu, Y. Xu, P. J. Shi, B. S. Xu, X. L. Wang, and Q. Liu, "Tribological properties and lubricating mechanisms of Cu nanoparticles in lubricant," Transactions of Nonferrous Metals Society of China, vol. 18, no. 3, pp. 636–641, 2008. · ·

88. B. Yu, Z. Liu, F. Zhou, W. Liu, and Y. Liang, "A novel lubricant additive based on carbon nanotubes for ionic liquids," Materials Letters, vol. 62, no. 17-18, pp. 2967–2969, 2008. · ·

89. B. Wang, X. Wang, W. Lou, and J. Hao, "Rheological and tribological properties of ionic liquid-based nanofluids containing functionalized multi-walled carbon nanotubes,"Journal of Physical Chemistry C, vol. 114, no. 19, pp. 8749–8754, 2010.

90. L. J. Wang, C. W. Guo, and R. Yamane, "Experimental research on tribological properties of $Mn_{0.78} Zn_{0.22} FE_2O_4$ magnetic fluids," Journal of Tribology, vol. 130, no. 3, Article ID 031801, 2008. ·

91. S. Chen and D. H. Mao, "Study on dispersion stability and self-repair principle of ultrafine-tungsten disulfide particulates," Advanced Tribology, vol. 995, 2010.

92. D. X. Peng, C. H. Chen, Y. Kang, Y. P. Chang, and S. Y. Chang, "Size effects of SiO_2 nanoparticles as oil additives on tribology of lubricant," Industrial Lubrication and Tribology, vol. 62, no. 2, pp. 111–120, 2010. ·

93. L. Vékás, D. Bica, and M. V. Avdeev, "Magnetic nanoparticles and concentrated magnetic nanofluids: synthesis, properties and some applications," China Particuology, vol. 5, no. 1-2, pp. 43–49, 2007. ·

94. R. E. Rosensweig, "Magnetic fluids," Annual Review of Fluid Mechanics, vol. 19, pp. 437–463, 1987.

95. Y. S. Kim, K. Nakatsuka, T. Fujita, and T. Atarashi, "Application of hydrophilic magnetic fluid to oil seal," Journal of Magnetism and Magnetic Materials, vol. 201, no. 1–3, pp. 361–363, 1999.

96. Y. Mitamura, S. Arioka, D. Sakota, K. Sekine, and M. Azegami, "Application of a magnetic fluid seal to rotary blood pumps," Journal of Physics Condensed Matter, vol. 20, no. 20, Article ID 204145, 2008. ·

97. Y. S. Kim and Y. H. Kim, "Application of ferro-cobalt magnetic fluid for oil sealing,"Journal of Magnetism and Magnetic Materials, vol. 267, no. 1, pp. 105–110, 2003. ·

98. L. Zhang, Y. Jiang, Y. Ding, M. Povey, and D. York, "Investigation into the antibacterial behaviour of suspensions of ZnO nanoparticles (ZnO nanofluids)," Journal of Nanoparticle Research, vol. 9, no. 3, pp. 479–489, 2007. · ·

99. R. Jalal, E. K. Goharshadi, M. Abareshi, M. Moosavi, A. Yousefi, and P. Nancarrow, "ZnO nanofluids: green synthesis, characterization, and antibacterial activity," Materials Chemistry and Physics, vol. 121, no. 1-2, pp. 198–201, 2010. · ·

100. N. Jones, B. Ray, K. T. Ranjit, and A. C. Manna, "Antibacterial activity of ZnO nanoparticle suspensions on a broad spectrum of microorganisms," FEMS Microbiology Letters, vol. 279, no. 1, pp. 71–76, 2008. · ·

101. Y. Liu, L. He, A. Mustapha, H. Li, Z. Q. Hu, and M. Lin, "Antibacterial activities of zinc oxide nanoparticles against Escherichia coli O157:H7," Journal of Applied Microbiology, vol. 107, no. 4, pp. 1193–1201, 2009. · ·

102. O. Mahapatra, M. Bhagat, C. Gopalakrishnan, and K. D. Arunachalam, "Ultrafine dispersed CuO nanoparticles and their antibacterial activity," Journal of Experimental Nanoscience, vol. 3, no. 3, pp. 185–193, 2008. · ·

103. P. Gajjar, B. Pettee, D. W. Britt, W. Huang, W. P. Johnson, and A. J. Anderson, "Antimicrobial activities of commercial nanoparticles against an environmental soil microbe, Pseudomonas putida KT2440," Journal of Biological Engineering, vol. 3, p. 9, 2009. · ·

104. H. J. Lee, S. Y. Yeo, and S. H. Jeong, "Antibacterial effect of nanosized silver colloidal solution on textile fabrics," Polymer Journal, vol. 8, p. 2199, 2003.

105. A. Paná ek, L. Kvítek, R. Prucek et al., "Silver colloid nanoparticles: synthesis, characterization, and their antibacterial activity," Journal of Physical Chemistry B, vol. 110, no. 33, pp. 16248–16253, 2006. ·

106. L. Brunet, D. Y. Lyon, E. M. Hotze, P. J. J. Alvarez, and M. R. Wiesner, "Comparative photoactivity and antibacterial properties of C60 fullerenes and titanium dioxide nanoparticles," Environmental Science and Technology, vol. 43, no. 12, pp. 4355–4360, 2009. ·

107. D. Y. Lyon and P. J. J. Alvarez, "Fullerene water suspension (nC_{60}) exerts antibacterial effects via ROS-independent protein oxidation," Environmental Science and Technology, vol. 42, no. 21, pp. 8127–8132, 2008. ·

108. A. Vonarbourg, C. Passirani, P. Saulnier, and J. P. Benoit, "Parameters influencing the stealthiness of colloidal drug delivery systems," Biomaterials, vol. 27, no. 24, pp. 4356–4373, 2006. · ·

109. R. Singh and J. W. Lillard, "Nanoparticle-based targeted drug delivery," Experimental and Molecular Pathology, vol. 86, no. 3, pp. 215–223, 2009. · ·

110. P. Ghosh, G. Han, M. De, C. K. Kim, and V. M. Rotello, "Gold nanoparticles in delivery applications," Advanced Drug Delivery Reviews, vol. 17, p. 1307, 2008.

111. M. Nakano, H. Matsuura, D. Ju, et al., "Drug delivery system using nano-magnetic fluid," in Proceedings of the 3rd International Conference on Innovative Computing, Information and Control (ICICIC ‹08), Dalian, China, June 2008.

112. A. Bianco, K. Kostarelos, and M. Prato, "Applications of carbon nanotubes in drug delivery," Current Opinion in Chemical Biology, vol. 9, no. 6, pp. 674–679, 2005.

113. C. Tripisciano, K. Kraemer, A. Taylor, and E. Borowiak-Palen, "Single-wall carbon nanotubes based anticancer drug delivery system," Chemical Physics Letters, vol. 478, no. 4–6, pp. 200–205, 2009. · ·

114. G. Pastorin, W. Wu, S. Wieckowski et al., "Double functionalisation of carbon nanotubes for multimodal drug delivery," Chemical Communications, no. 11, pp. 1182–1184, 2006. · ·

115. Z. Liu, X. Sun, N. Nakayama-Ratchford, and H. Dai, "Supramolecular chemistry on water-soluble carbon nanotubes for drug loading and delivery," ACS nano, vol. 1, no. 1, pp. 50–56, 2007. ·

116. X. Sun, Z. Liu, J. T. Robinson, et al., "Nano-graphene oxide for cellular imaging and drug delivery," Nano Research, vol. 1, p. 203, 2008.

117. L. Zhang, J. Xia, Q. Zhao, L. Liu, and Z. Zhang, "Functional graphene oxide as a nanocarrier for controlled loading and targeted delivery of mixed anticancer drugs," Small, vol. 6, no. 4, pp. 537–544, 2010. ·

118. Z. Liu, J. T. Robinson, X. Sun, and H. Dai, "PEGylated nanographene oxide for delivery of water-insoluble cancer drugs," Journal of the American Chemical Society, vol. 130, no. 33, pp. 10876–10877, 2008. · ·

119. X. Yang, X. Zhang, Y. Ma, Y. Huang, Y. Wang, and Y. Chen, "Superparamagnetic graphene oxide-Fe_3O_4 nanoparticles hybrid for controlled targeted drug carriers," Journal of Materials Chemistry, vol. 19, no. 18, pp. 2710–2714, 2009. ·

120. X. Fan, H. Chen, Y. Ding, P. K. Plucinski, and A. A. Lapkin, "Potential of ‹nanofluids› to further intensify microreactors," Green Chemistry, vol. 10, no. 6, pp. 670–677, 2008. · ·

121. M. J. Kao, C. H. Lo, T. T. Tsung, Y. Y. Wu, C. S. Jwo, and H. M. Lin, "Copper-oxide brake nanofluid manufactured using arc-submerged nanoparticle synthesis system," Journal of Alloys and Compounds, vol. 434-435, pp. 672–674, 2007. · ·

122. M. J. Kao, H. Chang, Y. Y. Wu, T. T. Tsung, and H. M. Lin, "Producing Aluminum-oxide brake nanofluids derived using plasma charging system," Journal of the Chinese Society of Mechanical Engineers, vol. 28, p. 123, 2007.

123. T. Sharma, A. L. M. Reddy, T. S. Chandra, and S. Ramaprabhu, "Development of carbon nanotubes and nanofluids based microbial fuel cell," International Journal of Hydrogen Energy, vol. 33, no. 22, pp. 6749–6754, 2008. · ·

124. J. Philip, T. Jaykumar, P. Kalyanasundaram, and B. Raj, "A tunable optical filter,"Measurement Science and Technology, vol. 14, no. 8, pp. 1289–1294, 2003.

125. A. Mishra, P. Tripathy, S. Ram, and H. J. Fecht, "Optical properties in nanofluids of gold nanoparticles in poly (vinylpyrrolidone)," Journal of Nanoscience and Nanotechnology, vol. 9, no. 7, pp. 4342–4347, 2009. · ·

126. W. Yu, H. Xie, and D. Bao, "Enhanced thermal conductivities of nanofluids containing graphene oxide nanosheets," Nanotechnology, vol. 21, no. 5, Article ID 055705, 2010. ·

127. W. Yu, H. Xie, and W. Chen, "Experimental investigation on thermal conductivity of nanofluids containing graphene oxide nanosheets," Journal of Applied Physics, vol. 107, no. 9, Article ID 094317, 2010. ·

128. Z. H. Han, F. Y. Cao, and B. Yang, "Synthesis and thermal characterization of phase-changeable indium/polyalphaolefin nanofluids," Applied Physics Letters, vol. 92, no. 24, Article ID 243104, 2008. ·

Emerging Trend in Natural Resource Utilization for Bioremediation of Oil – Based Drilling Wastes in Nigeria

Iheoma M. Adekunle[1], Augustine O. O. Igbuku[2],
Oke Oguns[3], and Philip D. Shekwolo[2]

[1]Environmental Remediation Research Group, Department of Chemical Sciences (Chemistry), Federal University Otuoke, Bayelsa State, Nigeria

[2]Restoration of Ogoniland Project Team, Shell Petroleum Development Company, Port Harcourt, Nigeria

[3]Remediation Team, Shell Petroleum Development Company, Port Harcourt, Nigeria

INTRODUCTION

Background

Nigeria is a country endowed with diverse mineral and natural resources among which is petroleum, a pivot to the national economy and sustainable development. In the past five decades, petroleum exploration and production activities have brought national economic boom but not without some aches. Acts of sabotage such as crude oil theft, pipeline bunkering and artisanal refining added to accidental spills and operational failures all combine to aggravate the oil-related aches. Oil spill into the environment, stemming from either acts of sabotage or operational failures, ultimately lead to environmental pollution with petroleum hydrocarbons [1, 2]. Petroleum mining or drilling is another factor to petroleum hydrocarbons in the environment. Most of the adverse impacts of oil spill/ petroleum hydrocarbons in the environment are experienced in the oil bearing communities, located in the Niger Delta region of the country; prominent among them being the Ogoni land pollution incidence reported by United Nations Environment Programme [1]. Petroleum exploration and production activities are strongly associated with drilling operations for oil mining. Accordingly, the extraction of petroleum resources from the earth is achieved by drilling activities. A developed drilling concept, irrespective of technological advancement, has its technical challenges, process requirements and environmental issues [3]. Drilling fluids, also referred to as drilling muds are used to enhance drilling activities via suspension of cuttings, pressure control, stabilization of exposed rocks, provision of buoyancy, cooling and lubricating.

Types of Drilling Fluids (Muds): There Are Basically Two Categories Of Drilling Fluids Namely (i) aqueous drilling muds or water based muds (WBMs), which consist of fresh or salt water containing a weighting agent, usually barite ($BaSO_4$), clay or organic polymers and various inorganic salts, inert solids, and organic additives to modify the physical properties of the mud so that it functions optimally and (ii) non-aqueous drilling fluids (NADFs), which comprise all non-water dispersible base fluids such as oil based muds (OBMs) and synthetic based muds (SBMs) [2]. Comparative evaluation of oil based muds and water based muds shows that OBMs offer advantages over WBMs for the reasons that [3]:

- OBMs are more suitable to drill sensitive shells, allowing drilling faster than the WBMs, providing excellent shale stability
- they are more adequate to drill formulations where bottom hole temperatures exceed WBMs tolerance, especially in the presence of contaminants such as water, gases, cement, salt and temperature up to 550F
- OBMs resist formation salt leach out
- they are characterized by thin filter cakes and the friction between the pipe and wellbore is minimized, thus, reducing the risk of differential sticking and are especially suited for highly deviated and horizontal wells
- the drill of low pore pressure formations is easily accomplished, since mud weight can be maintained at a weight less than that of water (as low as 7.5 ppg)
- corrosion of pipe is controlled since oil is the external phase and coats the pipe. The oils are non-conductors and the additives are thermally stable, hence, do not form corrosive products
- bacteria do not thrive long in OBMs
- there is the possibility of using OBMs over and over again and can be stored over long periods of time since bacterial growth is suppressed
- OBM packer fluids are designed to be stable over long periods of time even when exposed to high temperature and provide long-term stable packers since additives are extremely temperature stable. Properly designed, such packer fluids can suspend weighting materials over long periods of times.

In other words, regarding shale stability, penetration rate, high temperatures, drilling salts, lubrication, low pore pressure formations, corrosion control, re-use and packer fluids, OBMs offer advantages over WBMs. It is therefore, obvious that though WBMs are more environmentally benign, they are only satisfactory for less demanding drilling of conventional vertical wells at medium depths, whereas OBMs are more suited for greater depths or in directional or horizontal drillings, which exert greater stress on drilling apparatus. As a result, OBMs are more frequently used in petroleum industries for drilling purposes. The composition of OBMs include: petroleum base fluid, weighting agent and other chemical additives.

Drill Cuttings: During Drilling, Particles of Crushed Rocks Produced by the Grinding Action Of the drill bit as it penetrates the earth are referred to as drill cuttings (DC). DCs are, therefore, a mixture of rocks and particulates released from geological formulations in the drill holes made for crude oil drilling and are usually coated with the drilling fluid. Consequently, DCs are largely influenced by the chemical composition of drilling muds [2, 4].

The resultant spent OBM and drill cuttings (drilling wastes) consist of hydrocarbons, water, soils, heavy metals and water soluble salts such as chlorides and sulphates [3, 4]. Drilling wastes, which are toxic due to the presence of hydrocarbons, heavy metals and other chemical additives, if not properly treated before disposal, pose serious environmental hazards and risk to public health. Sequel to these, best practices in the management of drilling wastes cannot be over emphasized.

Health and Environmental Effects Associated with Drilling Wastes

Health effects linked to drilling wastes are traceable to the basic components such as the drilling fluid and additives:

Health Effects Associated with Drilling Fluids: These health effects are attributed to the physical and chemical properties of the drilling fluids. In oil based drilling wastes, the base oil stem from petroleum stream such as crude oil, diesel (gasoil) and kerosene, which cause skin irritation. Consequently, the most commonly observed health effect associated with drilling fluids is skin irritation. Other effects include headache, nausea, eye irritation and coughing. Routes of exposure in human are dermal, inhalation, oral and some other miscellaneous routes. On exposure to drilling fluid, petroleum hydrocarbons tend to remove natural fat from the skin, which results in skin drying and cracking. These conditions allow compounds to permeate through the skin leading to irritation and dermatitis. Susceptibility to these health effects varies with individual resistance capacity and conditions of poor personal/environmental hygiene. High aromatic content fluids, especially diesel fuel contain significant levels of carcinogenic polynuclear aromatic hydrocarbons (PAHs). Diesel fuels may also be genotoxic due to high proportions of 3-7 ring PAH [2]. Skin-painting studies in mice showed that, irrespective of the level of PAH, long-term dermal exposure to diesel fuels can cause skin tumours, an effect attributed to chronic skin irritation. In humans, chronic irritation may cause small areas of the skin to thicken, eventually forming rough wart-like growths, which may become malignant. Health effects from chronic exposure to PAHs may include cataracts, kidney damage, liver damage and jaundice. Naphthalene, a specific PAH, can cause the breakdown of red blood cells, if inhaled or ingested in large amounts. Animals exposed to levels of some PAHs over long periods in laboratory studies, developed lung cancer from inhalation and stomach cancer from ingesting PAHs in food [2].

Other hydrocarbon constituents of drilling fluids are the mono-aromatics popularly referred to as BTEX (benzene, toluene, ethylbenzene and xylene). BTEX compounds are very volatile, hence, will readily evaporate in warm/hot climates of tropical regions, resulting in higher concentrations in the vapor phase. As a result, there is the possibility of exposure to human via inhalation. Exposure to high concentrations of these hydrocarbons via inhalation may result in hydrocarbon induced neurotoxicity, a non-specific effect resulting in headache, nausea, dizziness, fatigue, lack of coordination, problems with attention and memory, gait disturbances and narcosis [2].

Health Effects Associated with Additives: In addition to the irritancy of the drilling fluid hydrocarbon constituents, several drilling fluid additives may also have irritant, corrosive or sensitizing properties. Various additives include emulsion stabilizers, pH adjusters, wetting agents, viscosifiers and fluid-loss reducing agents. For instance, calcium chloride ($CaCl_2$) has irritant properties and emulsifiers (such as polyamine) have been associated with sensitizing properties [3]. Specific chemical additives vary with locations.

Environmental Effects Associated with Drilling Wastes

Apart from health effects, environmental hazards associated with drilling wastes include land, water and air pollution [5]:

- *Land pollution:* Farming is the major land use system in Nigeria, especially in the Niger Delta region [1]. The most significant in this aspect of environmental pollution in Nigeria is thus farmland pollution. Consequences include alteration in soil physical, biological and chemical properties, loss of soil fertility, stunted plant growth and reduced crop productivity. These lead to reduced food security and compromised food safety.

- *Aquatic pollution:* Large percentage of the oil spill gets spread over the surface of the aquatic system resulting in anaerobic environment in the water, below the surface. This leads to death of the natural flora and fauna where oxygen is the key element for their respiration; adversely affecting fishing profession [1]

- *Air pollution:* volatile organics such as benzene, toluene, ethylbenzene and xylene could have elevated concentrations in the air, leading to atmospheric pollution and consequent adverse environmental and health impacts.

Oil well drilling processes generate large volumes of drill cuttings and spent mud in the country. Drilling wastes, therefore, add to hazardous petroleum waste materials released in the environments of the Niger Delta region of the country [1, 6] and the management of drilling wastes is quite tasking. An environmentally friendly technique for the management of drilling wastes is necessary in all offshore and onshore operations; from seismic surveys, drilling operations, field development and production to decommissioning. The physical and chemical properties of the drilling wastes influence their hazardous characteristics and environmental impact abilities, which in turn depend primarily on: (i) nature of impacted material, (ii) concentration of pollutant /amount of waste material after release (iii) recipient biotic community and (iv) exposure duration. Exposure that causes an immediate effect is called acute exposure while long-term exposure is called chronic exposure. Either acute or chronic exposure has negative impacts.

Contemporary Treatment of Drilling Waste Materials

Worldwide, contemporary drilling waste management options include re-use, offshore discharge, re-injection and onshore treatment and/or disposal [7]. Each treatment and or disposal option has its pros and cons as highlighted in the options (thermal technologies and bioremediation techniques) discussed.

Thermal Treatment

As the name suggests, thermal technologies involve the use of high temperatures to reclaim hydrocarbon contaminated materials [8]. Thermal treatment is mostly used in treating organic compounds. Additional treatment may be necessary for metals and salts depending on the final fate of the wastes. Thermal treatment technologies are designed for a fixed land based installation; however, a few mobile units exist. Two commonly practiced thermal treatment technologies are thermal desorption and incineration methods.

Thermal Desorption Method

Thermal desorption is an environmental remediation process that uses heat to increase the volatility of contaminants by the use of a series of equipment (desorber and oxidizer) such that the hydrocarbons and water are separated or removed from the solid matrix. It is normally carried out between the temperature range of 250-650°C. At these temperatures both the lighter and heavier hydrocarbons are removed and collected or thermally oxidized by further heating to a temperature of over 850°C. The resulting solid residue has essentially no residual hydrocarbons (having been oxidized), but does concentrate salts and heavy metals. Depending upon the success of process used, recovered hydrocarbons can be used as fuel or re-used as base fluid in the drilling fluid system and the resulting solid can be disposed of in a landfill or may be used in construction (of roads and bricks). Economical, operational and environmental implications of thermal desorption include:

- Effective removal and recovery of hydrocarbons from solids
- Possibility of recovering base fluid and end - product could be used for brick making
- Low potential for future liability
- Requires short time
- High cost of handling environmental issues

- Large volume of wastes is required to justify the cost of operation
- Requires tightly controlled process parameters
- High operating temperatures can lead to safety risks
- Requires several operators
- Heavy metals and salts are concentrated in residual solids
- Process water contains some emulsified oil
- Residue ash requires further treatment before disposal
- End product is sterile and can no longer support plant Life.

Incineration Method

Incineration involves (i) heating oil based mud and drill cuttings to a higher temperature range (1200-1500°C) in direct contact with combustion gases and (ii) oxidizing the hydrocarbons [8]. Solid/ash and vapor phases are generated. The gases produced from this operation may be passed through an oxidizer, wet scrubber, and bag house before being vented to the atmosphere. Stabilization of residual materials may be required prior to disposal to prevent constituents from leaching into the environment. Incineration of drilling wastes occurs in rotary kilns, which incinerate any waste regardless of size and composition. Incineration systems are designed to destroy only organic components of waste; however, most drilling wastes are non-exclusive in their content and therefore will contain both combustible organics and non-combustible inorganic materials. By destroying the organic fraction and converting it to carbon (IV) oxide and water vapor, incineration reduces waste volume. Inorganic components of wastes fed to an incinerator cannot be destroyed, only oxidized. The major inorganic materials are chemically classified as metals. Generally, these metals will exit the combustion process as oxides of the metals that enter. Economical, operational and environmental implications of incineration are as listed:

- Low potential for future liability
- High cost per volume
- Heat produced could be used for energy generation
- High energy cost
- Requires air pollution control equipment because of safety concerns

- At high temperatures, salts can form acid components
- Air emissions pose environmental concerns.

In line with best practices, for thermal technologies, there is need for proper placement of end product. Demonstration of sufficient compliance with current regulations and adequate safety measures to cater for the potential risks of exposure to high temperatures.

Bioremediation Technique

Bioremediation technique relies on the ability of microorganisms (mostly combination of bacteria) to feed on the hydrocarbons (HCs) as substrate, converting them into carbon dioxide, water and harmless clean solids; and the ability of some of the HCs to biodegrade over time. But in most cases, the native microorganisms are often overwhelmed by the extent of the hydrocarbon contamination and thus would require external nutrients to boost (bio-stimulation) their activity and ability to take up the HCs at a faster rate. In other cases, the native microorganisms may be needing help from their kind or other species of micro-organism which are grown or inoculated (bio-augmentation) in the laboratory and then introduced in the habitat of the native micro-organisms. Bioremediation could be carried out at the site of contamination (in-situ bioremediation technique) or off the site of contamination (ex-situ bioremediation technique). Bioremediation technologies include land farming, use of bioreactors, biopiles and compost- based technologies. Economical, operational and environmental implications of conventional bioremediation technique [9, 10, 11, 12, 13, 14] include:

- Relatively inexpensive
- Requires simple equipments and eliminates transportation cost as drill wastes could be treated on site
- Less capital but may be labour-intensive.
- Low maintenance cost; being a simple technology process that requires few machines, there are few delays due to equipment down-time
- Process is fairly flexible and can be used for most drill wastes including OBM, NADFs, previously extracted materials and newly drilled cuttings

- Proven technology
- Requires a considerable period of time to complete a process
- Appropriate bacteria and nutrient selection could be a daunting task
- In cases where bacteria are inoculated and brought on site, adaptability to their new environment may hamper their performance
- Minimal operation hazards
- Environmentally friendly: once the contaminants have been degraded, the microbial population reduces considerably as they have used up their food source
- Less impact on the environment as residue from process (TPH < 1%) may require no further treatment and could be used for agricultural purposes.

Recommended best practices for bioremediation technology include ensuring (i) proper initial physical, biological and chemical characterizations to determine extent of organic and inorganic contamination, (ii) required skill and persistence for the selection of several combinations of bacteria and nutrients that can provide the desired result (iii) proper periodic tillage to provide for proper aeration that facilitates degradation of the HCs and (iv) an accurate and appropriate TPH level check in between treatment process in order to monitor progress of the remediation process. Choice of waste management options typically considers local regulations, environmental assessment, cost/benefit analysis and the composition of the drilling wastes. The Department of Petroleum Resources [15] via the Environmental Guidelines and Standards for the Petroleum Industry in Nigeria (EGSPIN) stipulated guidelines on drill cuttings discharge for inland / near-shore and offshore deep water in order to minimize the adverse impact on the surrounding environment. These requirements call for an appropriate drill cuttings treatment prior to disposal in order to meet the stipulated conditions.

Review of Emerging Trend in the Treatment of Drilling Waste Materials in Nigeria

There are scientific evidences showing that drilling wastes generated in the country contain toxicants that are of environmental concerns. For instance, the reports of [16] on the determination of selected physical and chemical parameters including metals concentrations in a certain drill cutting dump site in the country. Results from their study showed that oil and grease on the surface and 20 feet around the waste dump area were above the specified limit [15]. There was also lack of plant growth noticed in the study, attributed to depletion of nitrogen, phosphorus and potassium values below threshold levels for plant growth. The reports of [4] on hydrocarbon and some metal contents of drilling muds and cuttings generated during the drilling of Igbokoda onshore oil wells gave total petroleum hydrocarbon (TPH), aliphatic hydrocarbon (AH) and polycyclic aromatic hydrocarbon (PAH) as generally exceeding stipulated limits by both national and international agencies. The studies of [17] on the compositional distribution and sources of polynuclear aromatic hydrocarbons (PAHs) in Nigerian oil-based drill-cuttings, showed that the total initial PAHs concentration of the drill cuttings was 223.52 mg/kg while the initial individual PAHs concentrations ranged from 1.67 to 70.7 mg/kg, dry weight, with a 90% predominance of the combustion-specific 3-ring PAHs.

The commonly employed remediation techniques for drilling wastes in Nigeria appear to be thermal technologies. However, due to economical, operational and environmental implications of these thermal technologies; search for more acceptable techniques commenced. There is scarcity of literature on the use of natural resource materials for the remediation of drilling wastes in Nigeria. The few literature resources showed that a large percentage is still at the bench-scale platform. For instance, [18] isolated*Staphylococcus sp.* from oil-contaminated soil that was treated with 1% drilling fluid base oil (HDF-2000). Their study revealed that *Staphylococcus sp.,* is a strong primary utilizer of the base oil and has potential for application in bioremediation processes involving oil-based drilling fluids. On the other hand, the effectiveness of 2 bacterial isolates (*Bacillus subtilis* and *Pseudomonas aeruginosa*) in the restoration of oil-field drill-cuttings contaminated with polynuclear aromatic hydrocarbons was studied by [19]. In that study, a mixture of 4 kg of drill cuttings and 0.67 kg of top-soil were fed into triplicate plastic reactors labeled A1 to A3, B1 to B3, C1 to C3 and O1 to O3. These were left quiescent for 7 days under ambient conditions, followed by the addition of 20 mL working solution of pure cultures of *Bacillus* sp and *Pseudomonas* sp (each of cell density 7.6 x 10^{11} cfu/mL) to reactors A1 - A3 and B1 - B3 respectively. Another 20 mL working solution containing both cultures at cell density 1.5 x 10^{12} cfu/mL was added to reactors C1 - C3. The working solution was added to each reactor (excluding the controls, O1 - O3) every 2 weeks. Mixing and watering of the set-ups were carried out at 3 days interval under ambient temperature of 30°C for a period of 6 weeks. Results showed that the predominant 3-ring PAHs, which made up 90% w/w of the total PAHs concentration of 223.52 mg/kg, were degraded below detection and the 4-ring PAHs were reduced from 4 to 0.6% by *Pseudomonas* while *Bacillus* reduced 3 and 4-ring PAHs respectively to 0.2 and 0.8%. Their works revealed that Pseudomonas degraded 3 and 4-ring PAHs relatively better than *Bacillus*. Both strains of bacteria degraded 5 and 6-ring PAHs below detection limits. Furthermore within the 3-ring PAHs, each of the strains of bacteria reduced phenanthrene to approximately 0.2%, whereas both degraded homologues acenaphthylene, acenaphthene and fluorene as well as anthracene below detection limits. For 4-ring PAHs, *Pseudomonas* degraded fluoranthene and benzo[a]anthracene. *Bacillus* also degraded benzo[a]anthracene below detection limits. *Pseudomonas* was able to reduce pyrene and chrysene to 0.3 and 0.2% respectively; whereas *Bacillus* reduced fluoranthene, pyrene and chrysene to 0.1, 0.01 and 0.4% respectively. However, treatment with the mixed culture resulted in limited degradation of 5-ring PAHs particularly in the fourth week, which was attributed to the phenomena of co-metabolism and inhibition.

The works of [20] compared the potentials of bio-augmentation and conventional composting as bioremediation technologies for the removal of PAHs from oil-field drill-cuttings. From a mud-pit, close to a just-completed crude-oil well in the Niger Delta region of Nigeria, 4000 g of drill cuttings was obtained and homogenized with 667 g of top-soil (to serve as microbes carrier) in three separate reactors (A, B and C). The bio-augmentation of indigenous bacteria in the mix was done by adding to reactors A and B a 20-mL working solution (containing 7.6x10^{11} cfu/mL) of pure culture of *Bacillus* and *Pseudomonas*, respectively, while a 20-mL working solution (containing 1.5×10^{12} cfu/mL) of the mixed culture of *Bacillus* and *Pseudomonas* was added to reactor C. The bio-preparation was added to each reactor (excluding the control) every two weeks for six weeks. The composting experiment was conducted in a 10-litre reactor in which 4000 g of drill cuttings, 920 g of topsoil and 154 g of farmyard manure and poultry droppings were homogenized. Mixing and watering of the set-ups were carried out at 3 days interval under ambient temperature over a period of six weeks. Results showed that initial individual PAHs concentrations in the drill cuttings ranged from 1.67 to 70.7 mg/kg dry weight, with a predominance of combustion-specific 3-ring PAHs (representing 90% of a total initial PAHs. After the bioremediation exercise that lasted for 42 days, total PAHs in the drill cuttings were reduced from 223.52 to 4.25 mg/kg, representing a 98.1% reduction. Away from the use of microbial strains in the treatment of drilling wastes, a bench-scale investigation was carried out by [21] to demonstrate the efficacy of technique referred to as 'Dispersion by Chemical Reaction (DCR) technology". This particular method involved the use of hydrophobized calcium oxide (CaO) to form a dry, soil-like material that could be useful in construction works.

On the other hand, after the study on the response of four phytoplankton species in some sections of Nigeria coastal waters to crude oil in controlled ecosystem [22], that revealed the adverse impacts; a multidisciplinary environmental remediation research group (ERRG) was inaugurated with the mandate to embark on innovative, cutting-edge research and development (R & D) initiative, aimed at the development of an indigenous technology for an eco-friendly technique in the treatment of soils, sediments, sludge and drilling wastes polluted by petroleum hydrocarbons, using natural products of Nigeria origin. The goal of ERRG is to translate the technology from bench-scale to field scale and come out with on- the - shelf products that will find use for both onshore and offshore remediation works. The first phase of the R & D initiative was the exploration of the remediation potential of conventional composting technology based on the results from the works of [23]. A good start was the production of a scientifically formulated and classified compost bulk [24] that are potentially viable for environmental remediation projects [25] and able to biodegrade petroleum hydrocarbons embedded in soil and related matrices [26]. The next phase was to assess public acceptance of the principles of this technology, which culminated to the reports of [27] on population perception impact on value-added solid waste disposal in developing countries, a case study of Port Harcourt City. The feedstock utilized in product formulations in this emerging, indigenous and innovative technology is 100% biodegradable and very abundant in the Nigerian environment. Consequently, the technology has been categorized by stakeholders [27] as:

- eco-friendly environmental remediation technique
- waste to wealth initiative
- waste to resource initiative
- value-added waste management option
- a contribution to the promotion of local material development that has the potential for:
- wealth creation
- job creation
- poverty alleviation
- sound environmental management of hydrocarbon polluted wastes from the petroleum industries.

ERRG observed that either conventional composting technology or bioremediation via utilization of pure microbial isolates/strains has limitations in terms of serving the practical needs of the petroleum industry in Nigeria with regards to meeting (i) regulatory remediation targets at close – out of project and (ii) project delivery time. Subsequently, through series of bench-scale and screen house remediation investigations, products were formulated to enhance the speed of bioremediation process using nano-scale green catalysts, a technique that matured into Compost - based Nanotechnology in Bioremediation (CNB-Tech). The research group then subjected the CNB-Tech products to different scientific evaluations in order to ascertain (i) efficiency on biodegradation of petroleum hydrocarbons in oily wastes such as crude oil impacted soils, sludge and drilling wastes (drill cuttings and oil-based mud) and (ii) environmental impacts with emphasis on soil quality. Published works on assessment and prognosis of products' impact on soil quality include:

- Assessing the effect of bioremediation agent from local resource materials in Nigeria on soil pH [28]

- Impact of bioremediation formulation from Nigeria local resource materials on moisture contents for soils contaminated with petroleum [29]
- Assessing and forecasting the impact of bioremediation product derived from Nigeria local raw materials on electrical conductivity of soils contaminated with petroleum products [30]
- Soil temperature dynamics during bioremediation of petroleum products using remediation agent from Nigerian local resource materials [31].

Other works on CNB-Tech products' evaluations including (i) effect on soil heavy metal dynamics and (ii) impact on soil microbial species population and diversity are being considered elsewhere for publication. Having recorded a huge success during the laboratory scale investigations where maximum of 4000g of sample bulk and freshly hydrocarbon contaminated soils (similar to the quantities used by other investigators) [19, 20] were treated, it became necessary to assess the efficiency of CNB-Tech products on waste materials with complex nature and higher degree of hydrocarbon pollution. This aspiration was realized in collaboration with the Remediation Department of Shell Petroleum Development Company (SPDC), Port Harcourt, Nigeria through the University Liaison Team of SPDC. Sequel to this, pilot-scale projects were commissioned to evaluate the efficiency of CNB-Tech products on the degradation of hydrocarbon compounds in the following petroleum impacted materials:

Hydrocarbon polluted clay soils from Ejama-Ebubu legacy site of SPDC

- Hydrocarbon polluted carbonized soil from Ejama-Ebubu legacy site of SPDC
- Hydrocarbon polluted sludge from Ejama-Ebubu legacy site of SPDC
- Oil-based mud and drill cuttings generated from SPDC operations.

Ejama Ebubu is one of SPDC's legacy sites of up to 42 year long pollution as at the time of study in 2011 [1]. In this chapter, the efficacy of CNB-Tech products in the biodegradation of petroleum hydrocarbons in oil-based drilling wastes (OBM-DC) is presented.

Research Justification

The treatment of drilling wastes, especially OBM-DC in an environmentally sound manner is a challenging task due to the complex nature of the wastes. The most popular technique adopted for the treatment of OBM-DC, thermal desorption [15] has its accompanying environmental concerns. For instance, thermal treatment technologies are associated with prohibitive capital and operational cost implications, threatening environmental consequences in addition to high occupational hazards and generation of secondary waste stream that has to be treated at extra high cost before final disposal. Consequently, there is need for a pragmatic shift to seek alternative techniques that will address the need of the oil and gas sector in the management of drilling wastes in terms of remediation target delivery time and compliance to regulatory standards in Nigeria. Regulatory standards for close-out of remediation projects vary from one country to another and success factors of a given technology are dependent on indices such as:

- climatic conditions
- geographical characteristics of the location
- nature and complexity of contamination
- expected utility of the end-products of the remediation exercise

It then becomes evident that a successful remediation technology in one part of the globe may not necessarily be efficient in another region, pointing to the need to look inward for a more practical approach to solving the environmental challenges posed by petroleum hydrocarbon polluted waste streams in Nigeria [1]. Having run laboratory, bench-scale and screen-house remediation works using CNB-Tech products on fresh hydrocarbon contaminated soils, it became necessary to conduct pilot scale remediation works on more challenging waste streams such as weathered petroleum impacted soils, sludge, sediment, oil- based drilling mud and drill cuttings, hence this project.

Research Objectives

The current study comprised three major objectives:

- to conduct a review on the emerging trends in the treatment and related studies for drilling wastes in Nigeria,
- to assess the efficiency of an indigenous and innovative application of compost - based nanotechnology in bioremediation (CNB-Tech) in biodegradation of hydrocarbons found in oil-based mud and drill cuttings; generated by a petroleum industry in Nigeria
- to investigate the beneficial utility of the remediation end-product for agricultural purpose (crop production), which is a major land use system in Nigeria.

RESEARCH METHODOLOGY

The research methodologies employed in this study were:

- Literature review to provide an insight to the current and emerging trend in the treatment of drilling waste materials in the country and
- Practical, ex-situ, pilot scale execution of biodegradation of hydrocarbon compounds in oil-based mud and drill cuttings generated by an oil company in Nigeria using an indigenous and innovative biotechnological (CNB-Tech) approach anchored on the use of natural resource materials of Nigeria origin.

Pilot-Scale Remediation of Oil-Based Mud and Cuttings Using CNB-Tech Method

This study was carried out during the 2010/ 2011 Sabbatical Programme of the University Liaison Team of Shell Petroleum Development Company (SPDC); in conjunction with the Remediation Department of SPDC, Port-Harcourt, Nigeria. The indigenous remediation products (CNB-Tech products) prepared from cellulosic natural resource materials and biogenic nanopolymers of Nigeria origin used for this pilot remediation study, were denoted as (i) Ecorem, (ii) Bioprimer and

(iii) Biozator. The last two products are solids that are transformed to the aqueous form before use while the first product is used in the solid form.

Project Site Description

The present pilot-scale project, for the purposes of adequate monitoring and efficient execution, was carried out in the Industrial Area of Shell Petroleum Development Company, Port Harcourt, Rivers State; known as "Shell IA". The earmarked project area was a relatively isolated open green field within Shell IA and according to design, a temporary sheltered facility constructed to suit the project design was erected at the site and all necessary health and safety issues were taken into consideration. The sheltered project facility comprised of three major units:

- Remediation execution section: where actual remediation took place
- Phyto-analytical section: where effects on plant life were investigated
- Mini- chemical laboratory: where necessary onsite chemical evaluations were conducted.

Pilot Scale Remediation Procedure

The batch of oil-based mud and drill cuttings (OBM-DC) used in this study was generated from SPDC's operations and supplied by one of the company's certified vendors. During the conveyance procedure for OBM-DC, chain of custody document and waste stream tracking manifest was observed. Basic highlights for CNB-Tech application mode are outlined in Figure 1. Pretreatment involved recovery of free phase base fluid and stabilization involved modification of viscosity parameter.

In order to collect sample from a particular replicate, each replicate was subdivided into 4 equal parts; representative fractions were collected from the different parts and recombined to give a composite sample of 1kg.

BTEX Sampling: Standard sampling kit for BTEX, sent by RespirTEK Consulting Laboratory, was utilized for the purpose. In this procedure, homogenized samples were collected from the cells using "Terra Core" sampling device. Using a 40 mL glass VOA vial containing appropriate preservatives and with the plunger seated in the handle, the Terra Core was pushed into freshly homogenized sample until the sample chamber was filled to the capacity of 5g. All sample particulates (debris) were removed from the outside of the Terra Core sampler and the sample plug was pushed into the mouth of the sampler. Excess soil that extended beyond the mouth of the sampler was removed. The plunger was then seated in the handle and rotated until it aligned with the slots in the body. The mouth of the sampler was placed into the 40 mL VOA vial containing the preservatives and sample extruded by pushing the plunger down. The lid was quickly placed back on the 40 mL VOA vial. It was ensured that when capping the 40 mL VOA vial, sample debris was removed from the top of the vial.

All samples were appropriately labeled and recorded in the chain of custody form before shipping to the USA laboratory by courier. Two Laboratories in Nigeria also collected samples for analyses, following standard procedures. The third laboratory in Nigeria was only involved in the analysis of materials using infrared and UV-absorption spectroscopic methods.

Physicochemical Analysis and Microbial Assessment

Statement from quality control and quality assurance unit (QA/QC) of RespirTek Laboratory, USA showed that all analyses were conducted following procedures set forth by the ISO/IEC 17025:2005 accreditation program standards for which the laboratory holds certification. Quality assurance systems and quality control criteria were strictly followed. The following parameters were determined:

- Total petroleum hydrocarbons (TPH)
- Monoaromatic hydrocarbons: benzene, toluene, ethylbenzene and xylene (BTEX). For xylene, ortho -, meta - and para- derivatives were assessed
- PAHs: a total of 17 PAH compounds: (i) naphthalene, (ii) acenaphthylene, (iii) acenaphthene, (iv) fluorene, (v) phenanthrene, (vi) anthracene, (vii) fluoranthene, (viii) pyrene, (ix) benzo (a) pyrene, (x) chrysene, (xi) benzo (b) fluoranthene, (xii) benzo (k)fluoranthene, (xiii) benzo (a) pyrene, (xiv) dibenz(a,b) anthracene, (xv) benzo (ghi)perylene, (xvi) 2-methylnaphthalene and (xvii) indeno (1,2,3-cd) pyrene
- Metals: barium (Ba), calcium (Ca), copper (Cu), lead (Pb), mercury (Hg), Nickel (Ni), Sodium (Na), Potassium (K), cadmium (Cd), zinc (Zn) and arsenic (As), a metalloid
- Miscellaneous parameters: pH, salinity, nitrogen, phosphorus, total organic carbon and electrical conductivity.
- Microbial activity: assessment of 48 hr and 96 hr microbial activities of both remediation end-product and contaminated material (control) was conducted by the USA based laboratory. Total hydrocarbon utilizing bacteria as well as total microbial count were assessed by the Nigerian based laboratories.

Hydrocarbon compounds were analyzed using Gas chromatographic method, microbial assessment was carried out using heterotrophic plate count method and metals were determined using atomic absorption spectroscopic technique. All the other parameters were carried out using standard procedures such as described in [24, 25, 32]. The CNB-Tech products (Bioprimer and (Biozator) were characterized using infrared and UV-visible spectroscopic methods. The basic characteristics of Ecorem have already been reported in [24, 25] but was slightly enhanced, in this study, for case specificity.

Assessment of Seed Germination Potential of Treated Samples

The remediated materials used in this evaluation were not mixed with external soil and no external fertilizer material was added to the remediated soil. Seed germination potential (SGP) of treated samples were assessed and only viable maize seedlings were used for this purpose. In a remediated material matrix (4kg material contained in an experimental plastic pot), 6 seedlings of maize were sown. This was replicated three times. All together, 18 (6 x 3) seedlings were used to evaluate this effect. Similar set- ups were also established for the untreated oil – based mud and cuttings, which served as control systems. This gave a total of 18 (6 x 3) seeds tested for germination potential for the test systems and 18 seedlings for the control media. This phase of the evaluation lasted for 7 days.

Assessment of Process Fluid (Leachate) Effect on Plant Growth

Adequate leachate (process fluid) management strategy was put in place as leachate generated during remediation was recycled into the remediation process. However, this evaluation was to ensure or to prove that in the event of any leachate seepage there would be reduced environmental risk. This phytotoxicity assessment was carried out using a cereal (corn: Zea mays L.,) as an indicator crop and indices of toxicity were (i) root length and (ii) plant height. Experimental systems constituted of the following set-ups, where FS is dilution factor and SF stands for farm soil:

- Farm soil + tap water (Code: FS + water). This served as control system for (ii) and (iii)
- Farm soil + stock leachate (Code: FS + LDF-0). This served as control system for (iii)
- Farm soil + diluted leachate series:
- Farm soil + leachate DF-1 (Code: FS + LDF-1)
- Farm soil + leachate DF-2 (Code: FS + LDF-2)
- Farm soil + leachate DF-3 (Code: FS + LDF-3)
- Farm soil + leachate DF-4 (Code: FS + LDF-4)

For this assessment, bulk farm soil sample, obtained from a village (K-dere, part of Ogoniland) in Rivers State, was used. Soil was sieved through a mesh and transferred at 1.5 kg per pot and designated pots were treated to 70% approximate field capacity (determined against gravity) using equal volume of appropriate fluid (water, stock leachate or diluted leachate). The systems were allowed to stabilize for 2 weeks after which viable maize seedlings were sown at 3 per pot. As the plants grew, the soil systems were treated with equal volumes of the appropriate fluid to maintain appropriate moisture level, as required by plant. Experiment lasted for 2 weeks, at the end of which the heights were recorded and plants harvested. Caution was exercised to ensure that roots were not destroyed during harvest. Root lengths were then recorded and mean values per pot calculated for each parameter.

Evaluation of Beneficial Utilization of End-Product

Similar to the case in Section 2.4, in this evaluation, the remediated matrix was not mixed with any type of soil, neither was any external fertilizer administered. At close - out of the pilot-scale remediation project, the remediated materials were air dried, primed with one of CNB-Tech products (Ecorem) at a specified loading scheme and then utilized as a growth media. Primed end-products were transferred at 4 kg per pot of 4 liter capacity. Three indicator crops used for this project were:

- Corn (*Zea mays L.,*)
- Green leafy vegetable (Fluted pumpkin: *Telfairia Occidentalis*)
- Cassava (*Manihot esculenta Crantz*)

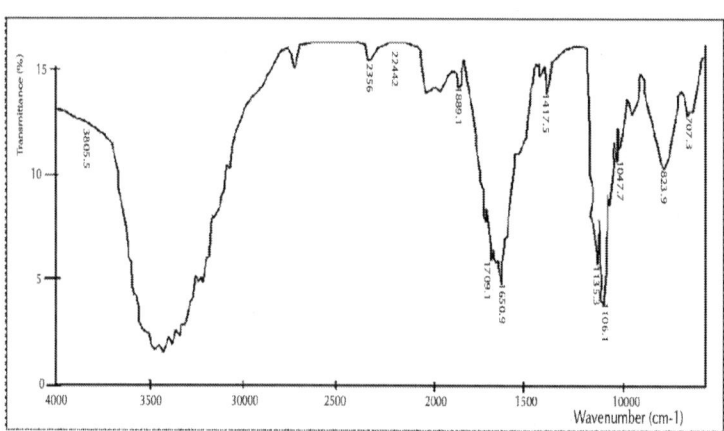

Figure 2: Infrared spectrum of Bioprimer, a CNB-Tech remediation product.

The crops were used because they are commonly grown and consumed in the Niger Delta region of the country. Due to time constraint, duration of investigation varied for the crops, the longest being up to 130 days for green leafy vegetable (Fluted pumpkin: *Telfairia Occidentalis*) while corn (*Zea mays L.,*) and cassava (*Manihot esculenta Crantz*) were grown for 2 and 3 weeks respectively. Untreated OBM-DC served as a control and farm soil served as a second control.

Statistical Analysis

Data generated in this study were subjected to statistical evaluations using SPSS software for Windows, version 17.0. Descriptive statistics were applied to evaluate mean and standard deviation. Paired sample T-Test and One-way analysis of variance (ANOVA) were applied to identify significant variations among treatments as appropriate. Pearson correlation was used to ascertain significant relationships.

RESULTS

Typical Infrared Spectra of two CNB- Tech Remediation Products

The infrared absorption spectra of two CNB-Tech products (Bioprimer and Biozator) utilized in this pilot scale study are presented in Figures 2 and 3. Both spectra showed absorption peaks in the region of 4000 to 600 cm^{-1}.

Major information from the infrared spectra were: strong, broad absorption band of oxygen-hydrogen (O-H) of an alcohol (aryl/ aliphatic) and N-H absorption bonds around 3500 - 3300 cm^{-1}; carbon-oxygen double bond (C=O) absorption band found around 1750 – 1500cm^{-1} This could be carbonyls of ester (RCOOR), aldehyde (RCHO), ketone (RCOR) and acid (RCOOH). C-N bond of nitrogenous matter falls in the end of the range; C-O bond around 1200 – 1000 cm^{-1} and of carbon-hydrogen (C-H) bond for aromatic moieties found below 1000cm^{-1} [33].

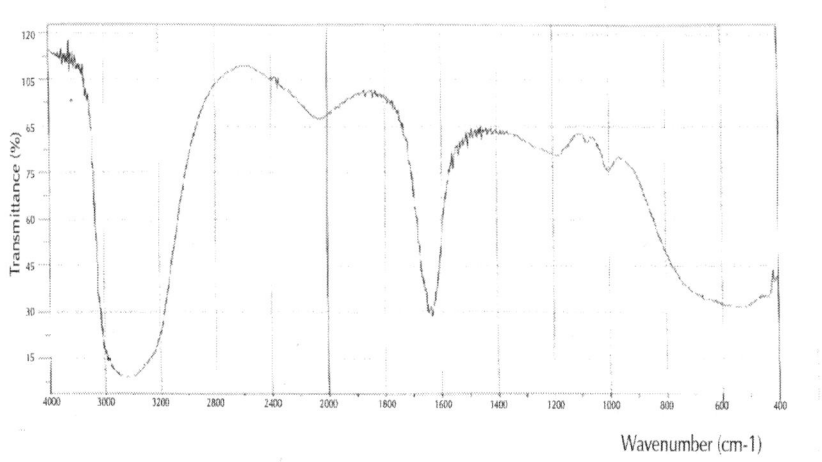

Wavenumber (cm-1)

Figure 3: Infrared spectrum of Biozator, a CNB-Tech remediation product.

Initial Characteristics of the Drilling Wastes

The results presented in this paper were largely those obtained from the International laboratory. Table 1 contains the initial characteristics of the drilling wastes (oil-based mud and cuttings).

Table 1: Initial characteristics of the oil -based drilling mud and cuttings used in this pilot scale study

S/N	Parameter	Concentration
Inorganics		
1.	Arsenic (mg/kg)	6.69
*2.	Cadmium	Not determined
3.	Barium(mg/kg)	765
4.	Calcium(mg/kg)	87300
5.	Copper(mg/kg)	35.90
6.	Lead(mg/kg)	161

7.	Mercury(mg/kg)	0.036
8.	Nickel(mg/kg)	12.3
9.	Sodium(mg/kg)	493
10.	Potassium(mg/kg)	1930
11.	Zinc(mg/kg)	144
12.	TKN (%)	0.0357
13.	Phosphorus (%)	0.0291
*14.	pH	10.2
*15.	Electrical conductivity (mSm^{-1})	Not determined
16	Total organic carbon (%)	Not determined
17..	Salinity (mg/kg)	4300
BTEX compounds		
1.	Benzene	0.0198
2.	Ethylbenzene	0.827
3.	m- and p-xylene	0.532
4.	o-xylene	0.924
5.	toluene	1.910
PAH Compounds		
1.	Naphthalene(mg/kg)	1.94
2.	Acenaphthylene(mg/kg)	BDL
3.	Acenaphthene(mg/kg)	BDL
4.	Fluorene(mg/kg)	2.54
5.	Phenanthrene(mg/kg)	0.78
6.	Anthracene(mg/kg)	BDL
7.	Fluoranthene(mg/kg)	BDL
8.	Pyrene(mg/kg)	BDL
9.	Benzo (a) anthracene(mg/kg)	BDL
10.	Chrysene(mg/kg)	BDL

11.	Benzo(b)fluoranthene(mg/kg)	BDL
12.	Benzo (k) fluoranthene(mg/kg)	BDL
13.	Benzo(a)pyrene(mg/kg)	BDL
14.	Dibenz(a,h) anthracene(mg/kg)	BDL
15.	Benzo(g,h)perylene(mg/kg)	BDL
16.	2-methylnapthalene(mg/kg)	5.39
17.	Indeno(1,23-cd) pyrene(mg/kg)	BDL
	Total PAH(mg/kg)	10.65
Total petroleum hydrocarbon		
1.	TPH (mg/kg)	79 200

[i] - *Parameters not determined by the USA laboratory but quantified by Nigerian based laboratories

Results indicated the presence of inorganic constituents and organics (hydrocarbons compounds). Regarding inorganics, soft metal contents increased in the order: Na (493 mg/kg) < K (1930 mg/Kg) < Ca (87, 300 mg/kg). The elemental ratios were 177 for Ca/Na, 45 for Ca/K and 4 for K/Na. Heavy metal concentrations increased in the order: Hg < As < Ni < Zn < Cu < Pb < Ba. In terms of hydrocarbon contents, total concentrations of polynuclear aromatic hydrocarbon (PAH) compounds was 10.65 mg/kg with concentrations of the individual components (Figure 4) increasing as phenanthrene (0.78 mg/Kg: 7%) < naphthalene (1.94 mg/kg; 18%) < fluorene (2.54mg/kg; 24%) < 2-methylnapthalene (5.39 mg/kg; 51%). Results on monoaromatics (BTEX), shown in Figure 5, gave a total concentration of 4.213 mg/kg out of which toluene constituted the highest fraction (45.34%), followed by xylene (34.56%), ethylbenzene (19.63%) and benzene (0.47%). Total xylene concentration was 1.456 mg/kg out of which ortho-xylene constituted 63.46% while meta- and para-xylenes gave 36.54% of the total (1.456 mg/kg).

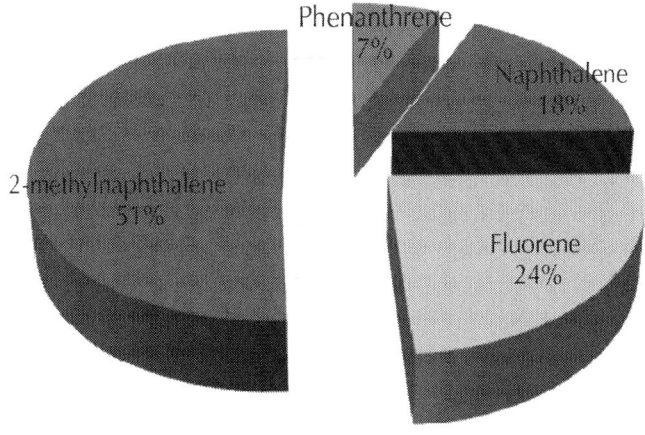

Figure 4: Percentage distribution of individual components of PAH relative to the total concentration.

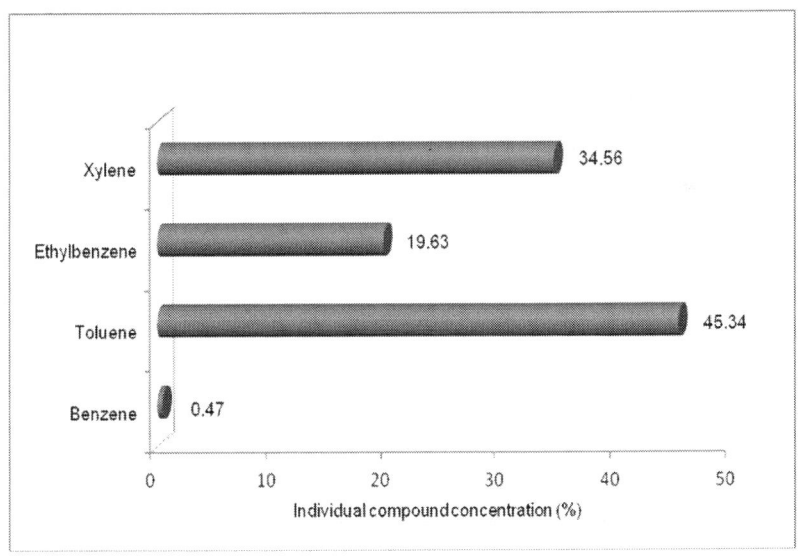

Figure 5: Percentage distribution of individual components relative to the total BTEX concentration.

Results on Petroleum Hydrocarbon Degradation

By application of CNB-Tech products, the initial TPH concentration of 79, 200 mg/kg decreased to 1888.67 ±161. 20 mg/kg. The difference in these two values was a mean TPH concentration of 77 311.33 ± 161.20 mg/kg. This difference corresponds to the total concentration of hydrocarbon compounds degraded or destroyed by the applied treatment. The initial concentration (79, 200 mg/kg) and the degraded fractions (in replicates of three) are presented in Figure 6. Specifically, results on hydrocarbon degradation (Figure 7) revealed 98% degradation for TPH, 100% degradation for BTEX and 100% degradation for PAH. Reduction in TPH level by 99% was obtained by the Nigerian laboratories.

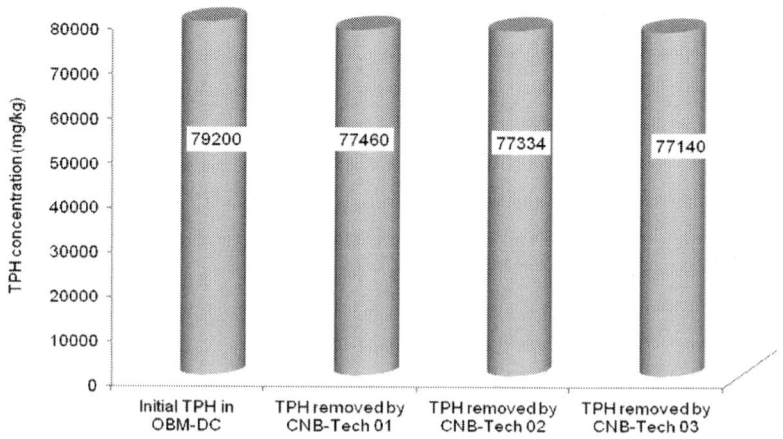

Figure 6: Graph showing concentrations of degraded TPH relative to the initial concentration.

Table 2: Qualitative results for the remediated media

S/N	Parameter	Remarks for contaminated medium	Remarks for remediated medium
1.	Appearance	Viscous, pasty and solid interfaced in oil suspension	Transformed to non-viscous, non-sticky crumby humus soil appearance
2.	Color	Light brown	Treated matrix had characteristic dark color of humus soil
3.	Odor	Presence of strong hydrocarbon odor	Complete disappearance of hydrocarbon odor in all the treated media and all treated samples exhibited clean earthy smell
4.	Sheen test	Strong oil sheen in water suspension	Complete disappearance of oil sheen in water suspension

Results on qualitative assessments of the untreated OBM-DC and remediated material in terms of appearance, odor, color and sheen test are contained in Table 2 and Figure 8 depicts the materials' appearances before and after remediation.

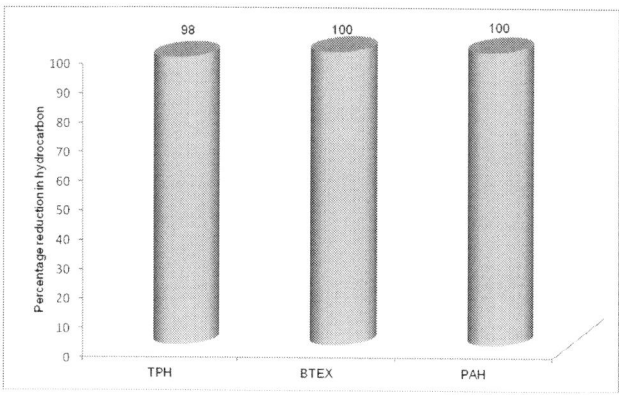

Figure 7: Percentage degradation of hydrocarbon compounds in the drilling wastes by applied CNB-Tech products.

Stalilzed OBM-DC before remediation

After remediation using produces

Figure 8: Photographs showing the materials before and after bioremediation by the application of CNB-Tech products.

Results on Inorganic Constituents of the CNB -Tech Treated Materials

Descriptive statistics of selected inorganic constituents found in the treated media are presented in Table 3. Changes in their concentrations relative to the initial values are presented in Figure 9. For instance, the initial pH value was reduced to 7.90 from 10.20, corresponding to 23% reduction. Likewise, the following reductions were obtained: 62% for Ca, 46% for As, 44% for Cu, 70% for Pb, 100% for Hg, 57% for Ni and 37% for Zn. The concentrations of some elements such as nitrogen, phosphorus and potassium were elevated. The nitrogen-phosphorus-potassium (NPK) status, as affected by treatment, is presented in Figure 10. Nigerian laboratories obtained the same trend for NPK status. Based on the results from USA, CNB-Tech remediation option applied in this study raised the nitrogen level from 0.036% to 0.096%, raised phosphorus level from 0.0291% to 0.312%, increased potassium by 1.4 fold (Figure 10) and sodium by 3 folds. The USA based laboratory did not analyze for total organic carbon and electrical conductivity but the Nigerian based laboratory did and recorded electrical conductivity in the range of 1956 to 2063 mSm^{-1} with a mean value of 2003 ± 54 mSm^{-1} before treatment. After remediation, the electrical conductivity of the end products ranged from 594 to 696 mSm^{-1} and a mean value of 640± 52 mSm^{-1}. From the mean values, there was a 68% reduction in electrical conductivity.

Table 3: Concentrations of some inorganic parameters in the treated materials

S/N	Element	Minimum	Maximum	Mean	Standard error	Standard deviation	Sample population
1.	pH	7.70	8.20	7.90	0.15	0.26	3
2.	Nitrogen (%)	0.070	0.130	0.096	0.016	0.028	3
3.	Phosphorus (%)	0.280	0.360	0.312	0.026	0.046	3
4.	Potassium (%)	0.50	0.77	0.61	0.08	0.14	3
5.	Copper (mg/kg)	18.10	21.70	20.10	1.06	1.83	3
6.	Zinc (mg/kg)	79.30	110	92.67	9.08	15.73	3
7.	Nickel (mg/kg)	3.99	7.05	5.29	0.92	1.59	3
8.	Calcium (mg/kg)	28900	39200	33466	3030	5248	3
9.	Arsenic (mg/kg)	2.50	4.85	3.59	0.68	1.18	3
10.	Lead (mg/kg)	5.87	54.80	27.06	14.50	25.12	3

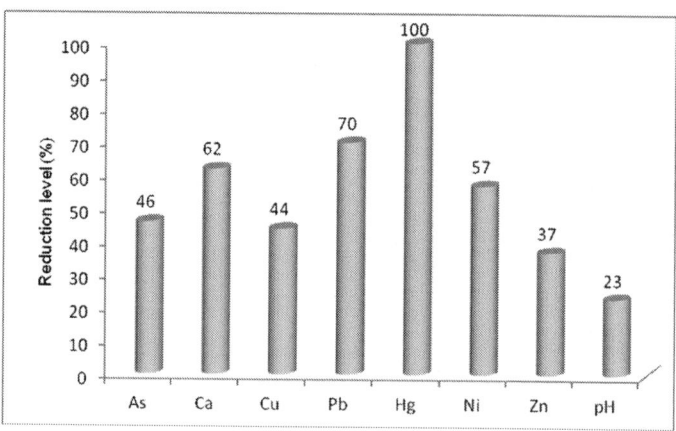

Figure 9: Reductions in some inorganic constituents of the drilling materials treated by CNB-Tech.

Total organic carbon ranged from 2.95 to 3.06% with a mean of 2.99± 0.06% before remediation and increased to 3.84 to 3.93% with a mean of 3.88 ± 0.05%; corresponding to an increase by 23%. Before remediation, Cd concentration varied from 6.70 to 7.60 mg/kg, with a mean value of 7.03± 0.49 mg/kg. After treatment, the metal concentration ranged from 0 to 1.80 mg/kg with an average of 1.05 ± 0.94 mg/kg. By the two mean values, cadmium level was reduced by 85% due to applied CNB-Tech products.

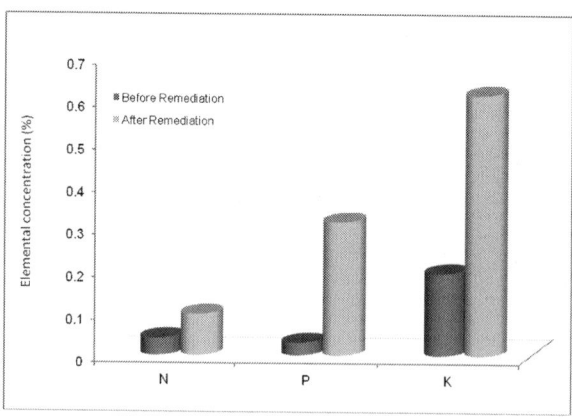

Figure 10: Nitrogen-phosphorus-potassium status before and after treatment as obtained by the USA based laboratory.

Results on Microbial Activity

The digital photographs of heterotrophic plate count results are shown in Figure 11. Microbial activities assessed on the untreated and treated samples revealed that the contaminated oil-based mud and cuttings (no. 1 in Figure 11), contained some indigenous microorganisms of up to 1.9×10^3 (cfu/mL) while the CNB-Tech remediated samples recorded up to a maximum of 3.15×10^7 cfu/mL. An illustration of microbial enumeration for 48-hr and 96 hr counts are presented in Figure 12.

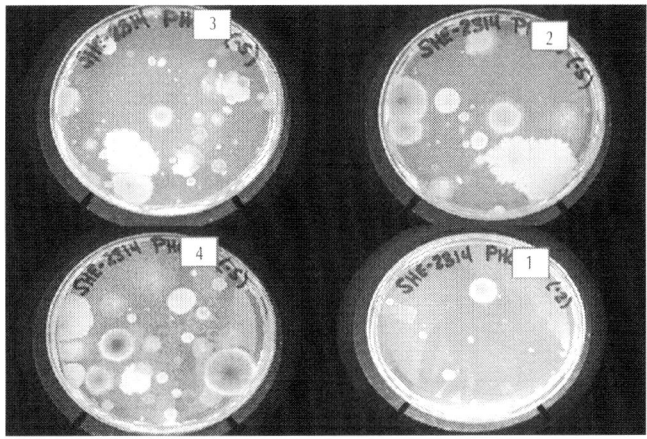

Figure 11: Heterotrophic plate count digital photographs for untreated OBM-DC (1) (before remediation) and replicates (2, 3, 4), after remediation using CNB-Tech method.

At 48 hr microbial activity assessment, maximum total microbial population of 1.9×10^3 cfu/mL was obtained for untreated OBM-DC and in the materials remediated by the application of CNB-Tech products, it was 1.45×10^7 cfu/mL. These two values were significantly different at $p \leq 0.05$. At 96 hr microbial activity assessment, a total microbial population of 2.4×10^3 cfu/mL was obtained for untreated OBM-DC and 3.15×10^7 cfu/mL for the remediated matrices. Results showed that within 48 hours, the microbial activity of the remediated matrices excelled over the untreated by over 7,000 folds and at 96 hours, it excelled by over 13, 000 folds, indicating rapid multiplication of microbial activity by CNB-Tech products which also increased with time.

Results on Phytotoxicity Assessment of Remediated Samples

Toxicity on Seed Germination Potential

The contaminated OBM-DC did not allow the germination of maize seedlings. Out of the sown 18 seedlings, none germinated. The untreated OBM-DC therefore, gave 100% toxicity to seed germination potential (SGP) of maize. On the contrary, all the 18 maize seedlings sown in the CNB-Tech remediated matrices germinated (Figure 13). Hence, resulting in 100% positive effect on SGP, indicating that the treated matrices exhibited 0% toxicity to seed germination.

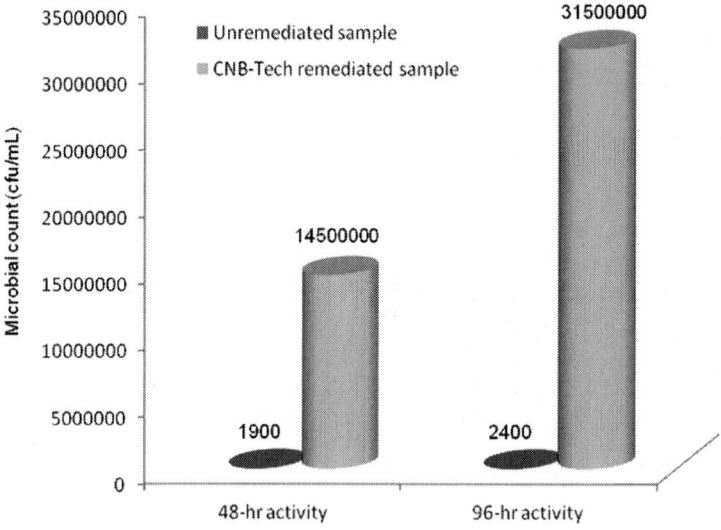

Figure 12: Microbial activity at 48 –hr and 96-hr counts for untreated oil-based drilling wastes and CNB-Tech remediated samples.

Figure 13: Germinated maize seedlings growing in treated media with picture taken on day 4 of growth.

Results on Beneficial Use of Remediation end Product

Figure 14, shows a cross-section of the treated materials (during recovery period) being aerated in preparation for use as plant growth media.

Figure 14: A cross section of project technical staff preparing the treated drilling wastes (OBM-DC) for use as plant growth media.

During the recovery phase of the remediated end-product, treated materials were allowed to lie fallow in order to establish natural processes as a sign of wellbeing and restoration. In this project, after the fallow period, early indications of material restoration were:

- spontaneous vegetative growth,
- the presence of larva within the spontaneously grown green vegetation,
- butterflies and small birds perching on the surface of the material, which could not take place before treatment

Remediated materials supported the growth of fluted pumpkin (*Telfairia occidentalis*). A cross-section of the green leafy vegetable at over 100 days of growth and that of cassava, at one week of growth, growing in the treated materials are shown in Figure 15. Narrowing to the height of *Telfairia occidentalis*, the mean height for crops grown in the untreated OBM-DC was 0 cm as there was complete inhibition to both germination and growth. The mean height for crops grown in CNB-Tech remediated media was 217± 25 cm, a value higher than the mean height (187± 40 cm) of the vegetable crops grown in farm soil collected from the region. The difference in the two mean values was significant at p = 0.14. Correlation for the heights of the vegetables grown in the treated media and those grown in the farm soil gave a coefficient of 0.95 (p = 0.204).

Fluted Pumpkin   Cassava

Figure 15: Remediated drilling wastes as plant growth medium for Fluted pumpkin (*Telfairia occidentalis*) and cassava (*Manihot esculenta Crantz*).

Results on the Impact of Remediation Leachate on Plant Life

Comparative evaluations of control system (soil treated with water only), stock leachate system (soil treated with leachate without any form of dilution) and systems treated with serial dilutions of the leachate (soil treated with leachate diluted with water by factors 1, 2, 3 and 4) are presented in Table 4.

Table 4: Impact of leachate generated at the close-out of project on the root length and height of maize

S/N	System Code	Leachate effect of on vegetative growth relative to control (%)		Effect of serial dilution on plant using stock (undiluted leachate) as reference (%)	
		Height	Root length	Height	Root length
1.	FS+ Water (Control)	Reference	Reference	Not applicable	Not applicable
2.	FS + DF-0	-1.50	-23.45	Reference	Reference
3.	FS + DF-1	32.60	1.12	34.62	32.20
4.	FS + DF-2	45.01	16.37	42.22	50.02
5.	FS + DF-3	66.86	21.37	69.41	58.55
6.	FS + DF- 4	75.39	24.51	78.07	62.66

[i] - Negative sign stands for decrease. The other positive values stand for increase, FS = farm soil and DF = dilution factor

Pictorial and graphical representations of leachate impact on plant height and root length are presented in Figures 16 and 17. Relative to the control system (soil treated with water only), leachate diluted with water by a factor of 4 improved plant height by 75.39% and root length by 24.51%. Figures16 and 17gave all the systems at a glance, relating the control (FS + Water), system SF+LDF-0 (DF-0) and serial dilutions (DF-1 = FS+ LDF-1, DF-2 = FS+ LDF-2, DF-3 = FS+ LDF-3 and DF-4 = FS + LDF - 4) for plant height and root length. Evaluating the effect of leachate dilution relative to the stock (undiluted) leachate, a 4-fold dilution excelled over the stock by 78.0% for plant height and 62.66% for root length. The relationships between plant height or root length and dilution factors are given in Figure 18. Pearson correlations gave strong coefficients: plant height versus dilution factor, r = 0.979 (p = 0.004), root length versus dilution factor, r = 0.932 (p = 0.021) and plant height versus root length, r = 0.972 (p = 0.006). From the results, plant vegetative growth increased with increasing dilution of leachate.

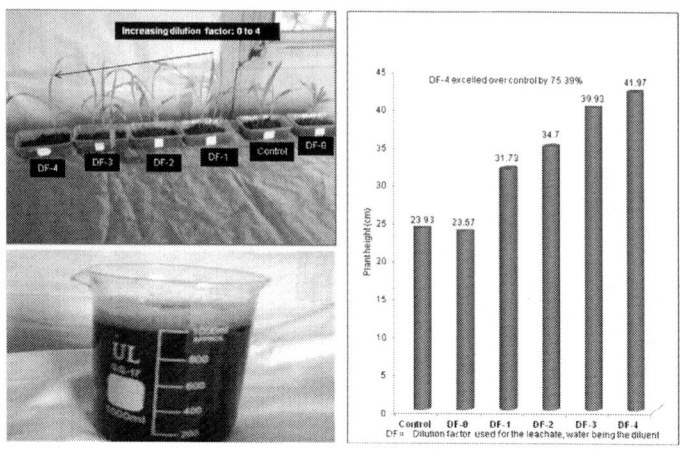

Figure 16: Pictorial and graphical representations of leachate impact on height of maize, including a picture of the stock leachate contained in a beaker.

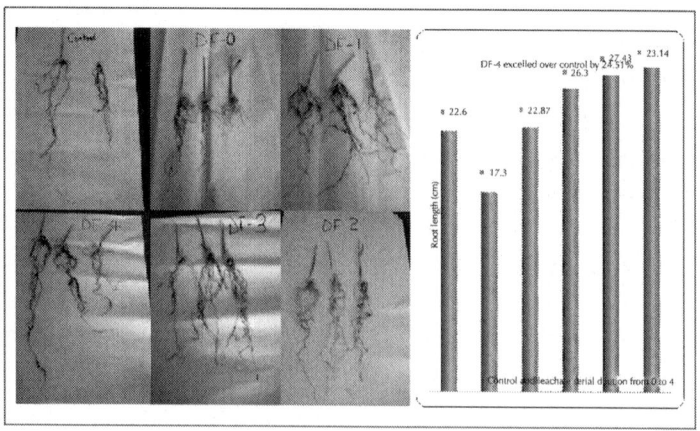

Figure 17: Pictorial and graphical representations of leachate impact on root length of maize.

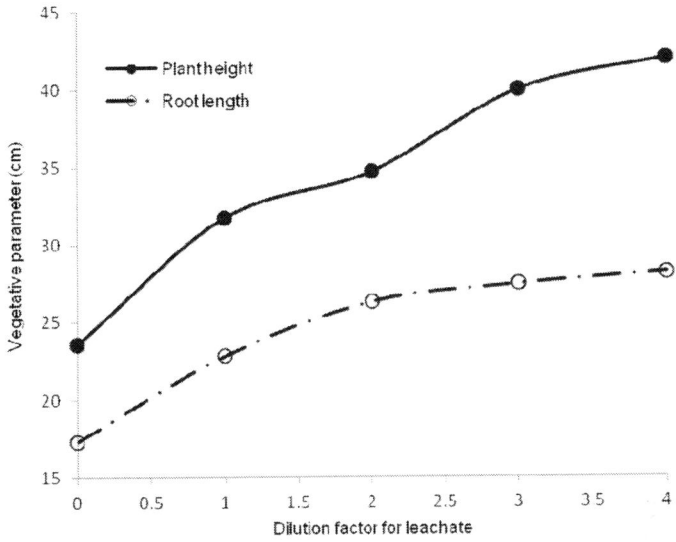

Figure 18: Relationship between plant vegetative growth and serial dilution of process fluid (leachate) generated during the remediation project.

DISCUSSION

The type of inorganic constituents and hydrocarbons found in the drilling wasting used in this study were consistent with the reports of [4, 17] but varied in concentrations. This confirms that the OBM-DC used in this study was toxic [2]. The remediation products of CNB-Tech series used in this study demonstrated a high (98 to 100%) degradation potential for the different constituents of hydrocarbon compounds found in the drilling wastes, within a short period of 6 days. This excellent performance was attributed to the chemistry, nature and operation mechanisms of the CNB-Tech formulations.

An infrared spectrum is primarily used to identify functional groups present in a molecular fragment [33]. The infrared spectra obtained for CNB-Tech products (Biozator and Bioprimer) revealed enrichment of the molecular structure of the two products with oxo- groups, indicating oxidizing functionality. The presence of C-H of aromatic nature and the O-H stretching absorption indicate the presence of both hydrophobic and hydrophilic properties, respectively, in their molecular fragments.

By implication, the remediation products are naturally endowed with:

- oxidizing ability
- polar (hydrophilic: water loving) molecular fragment
- non-polar fragment (hydrophobic: water insoluble, oil soluble) molecular fragment.

These natural endowments permit the dissolution of the products' active ingredients (solids) in water, making water the carrier medium for CNB-Tech liquid formulations. Consequently, Biozator and bioprimer are water based technical grade products. By the mentioned characteristics, the two products perform reduction and oxidation (Redox) reaction mechanisms, resulting in the degradation/ destruction of hydrocarbons compounds, without recombination to form new hydrocarbons. These absorption peaks in the infrared spectra further reveal that CNB – Tech products are natural hydrocarbon biodegradation catalysts for the following reasons:

- enhaced water solubility of hydrocarbons via sorption, hydrolysis and oxidation mechanisms
- enhanced bioavailability of hydrocarbon pollutants for microbial degradation
- increased supply of oxygen [O] molecules required for enhanced reduction –oxidation reactions in the hydrocarbon degradation process.
- surfactant property
- emulsification of hydrocarbons

The combined actions of hydrophobic molecular fragment, hydrolysis, oxidation and surfactant property of CNB-Tech products render hydrocarbons more water soluble and subsequently more available for biodegradation. Bioprimer and Biozator also emulsify hydrocarbons into droplets that can be easily assimilated by microorganisms. By these properties, the products reduce oil-water surface tension; enhance water solubility of petroleum hydrocarbons thereby enhancing the bioavailability of the contaminants (hydrocarbons) to microorganisms for both extracellular and intracellular decompositions. The two products are 100% biodegradable. The third CNB-Tech product used in this study (Ecorem: a black amorphous solid material, also 100% biodegradable) contains major and minor plant nutrient elements and via hydro-activation, naturally generates mixed consortia of microorganisms, which multiplies with time to facilitate the destruction of hydrocarbons. No engineered microorganism or externally imported microorganism was used in this study. This technology, therefore, saves time and eliminates the daunting task of isolating pure microbial strains and associated adaptability challenges linked with conventional bioremediation techniques [7, 8, 18, 19, 20].

The microorganisms from Ecorem product perform the following functions:

- extracellular decomposition in which the naturally produced microorganisms secrete enzymes to breakdown large organic compounds (such as hydrocarbons) into smaller forms for easier absorption into the micro-organisms. Once the smaller compounds have been absorbed by the microorganisms, intracellular decomposition takes place

- increased microbial activity facilitated by Ecorem, results in thermophilic temperature modulations in the range of 55 to 60°C, a process that accelerates degradation of hydrocarbons, especially polynuclear aromatic aromatic hydrocarbons (PAHs). Thermophilic temperature modulations also controls thermo-sensitive pathogen to crops animals and man; killing off weeds and seeds that will be detrimental to land use of end products.

By the above described mechanisms, the CNB-Tech products were able to biodegrade petroleum hydrocarbon compounds with high efficiency (98% degradation for TPH and 100% degradation for PAHs and BTEX) within a short period of time of 6 days, relative to previous works on bioremediation. For instance, in a study of in-situ bioremediation of oily sludge via biostimultaion of indigenous microbes, conducted by [34], through the addition of manure at the Shengli oilfield in Northern China for 360 days, 58.2% reduction in TPH was achieved in test plots and 15.5% reduction in control plot. By treating 2 kg of drill cuttings with initial TPH of 806.36 mg/kg for 56 days under the conditions of composting of spent oyster mushroom (*P.ostreatus*) substrate, [35] recorded overall degradation of PAHs in the range of 80.25 to 92.38%. In this present study, OBM-DC used had initial TPH of 79, 200 mg/kg and was degraded by 98% within the stated short period of 6 days. In a field trial biopile composting method [36] for drilling mud polluted sites in the Southeast of Mexico with comparable TPH level of 99 300 ± 23000 mg/kg, after 180 days, TPH concentrations decreased from 99 300 ± 23000 mg/Kg to 5500 ± 700 mg/kg, corresponding to 94% degradation for amended biopile and to 22900 ±7800 mg/kg, representing 77% decrease for unamended biopile. The mean residual value of TPH (5500 ± 700 mg/kg) left in the treated matrix in their study was higher than the mean residual value (1888± 161 mg/kg) obtained in this present study.

By conducting an investigation on two bioremediation technologies (bioremediation by augmentation and conventional composting using crude manure and straw) as treatment options for oily sludge and oil polluted soil in China [12] in which the total hydrocarbon content (THC) varied from 327.7 to 371.2 g/kg (327700 to 371200 mg/kg) for dry sludge and 151.0 g/kg (151000 mg/kg) for soil for a period of 56 days; after three times of bio-preparation application, THC decreased by 46 to 53% in the oily sludge and soil. The results (98 -100% degradation) obtained from this present study was from only one dose application of CNB-Tech products. Repeated application of CNB-Tech products by two to three dose applications will achieve 100% degradation of TPH. In another instance, a 5- month field scale bioremediation of sludge matrix via the utilization of organic matter such as bark chips via conventional composting, mineral oil (equivalent to total hydrocarbons) decreased from 2400 to 700 mg/kg (70% decrease) for sludge matrix and from 700 to 200 mg/kg, corresponding to 71% decrease [14]. In treating oil sludge using composting technology in semiarid conditions for 3 months, hydrocarbons were reduced from 250 to 300g/kg (250000 to 300 000 m/kg) by 60% against reduction by 32% recorded in the control [37]. The treatment applied by [37] and consequent reduction of 60% implies that the residual hydrocarbons in the treated samples would be between 100 000 and 180 000 mg/kg unlike the results obtained in this present study that gave residual hydrocarbon of 1888.67 ±161.20 mg/kg. In a study carried out by [38], sand samples contaminated with oil spill were collected from Pensacola beach (Gulf of Mexico) and tested to isolate fungal diversity associated with beach sands and investigate the ability of isolated fungi for crude oil biodegradation. From their results, 4.7 to 7.9% biodegradation was recorded.

Elsewhere in India, Abu Dhabi and Kuwait [39], bioremediation technology was applied in field-scale degradation of hydrocarbons in different oil wastes for a period of 12 months. Table 5 illustrates different reductions in total petroleum hydrocarbons obtained in these field case studies. TPH reductions in drilling wastes were obtained in the range of 90.85 to 95.48% with residual TPH in treated samples in the range of 2600 to 10 900 mg/kg (0.26 to 1.09%).

Table 5: Reductions in TPH levels obtained in field case studies of different types of petroleum impacted wastes (soils, drill cuttings and oil-based mud) in Abu Dhabi, Kuwait and India [39]

Name of the oil Installation / type of oily waste	Quantity of oily waste (cubic meter)	Number of batches	TPH Content (%) in oily waste before and after bioremediation		% Reduction in TPH	Residual TPH in treated material (%)
			Before	After		
Abu Dhabi National Oil Company (ADNOC), Abu Dhabi / Oil contaminated drill cuttings	200	1	17.26	0.98	94.32	0.98
BG Exploration and Production India Limited (BGEPIL), India / Oil based mud (OBM)	2,428	3	5.75 – 6.23	0.26 - 0.57	95.48-90.85	0.26 – 0.57
Bharat Petroleum Corporation Limited (BPCL), India / Oily sludge	5,000	1	19. 30 – 26.5	0.26 - 0.57	98.65-97.85	0.26 -0.57
Cairn Energy Pty. India Limited, India / Oil contaminated drill cuttings	567	2	14.93 – 18.81	0.82 – 1.09	94.51-94.21	1.09
Chennai Petroleum Corporation Limited (CPCL), India / Oily sludge	4,444	2	26.12	0.89	96.59	0.89
Hindustan Petroleum Corporation Limited (HPCL), India / Oily sludge	5,010	3	16.70 – 52.81	0.90 – 1.60	94.61-96.97	0.90-1.60
Indian Oil Corporation Limited (IOCL) Refineries in India / Oily sludge (acidic + non acidic)	75,412	48	9.6 – 38.4	0.37 – 0.95	96.15-97.53	0.37-0.95

Kuwait Oil Company (KOC), Kuwait / Oil contaminated soil	778	1	4.6 – 12.75	0.09 – 0.10	98.04-99.21	0.09-0.10
Mangalore Refinery and Petrochemicals Limited (MRPL), India. / Oily sludge	2,222	2	8.35 – 19.86	0.84 – 0.97	89.84-95.12	0.84-0.97
Oil and Natural Gas Corporation Limited (ONGC) installations in India / Oily sludge & oil contaminated soil	95,499	145	12.0 – 51.5	0.5 – 1.2	95.83-97.67	0.50--1.20
Oil India Limited (OIL) , Assam / Oily sludge & oil contaminated soil	15,921	14	21.6 – 37.7	0.49 – 0.53	97.73-98.59	0.49-0.53
Reliance Energy Limited (RIL), India / Oily sludge	611	2	19.15	0.5	97.39	0.50

The residual TPH level (1888.67 ± 161.20 mg/kg) obtained in this present study was below the Environmental Guidelines and standards for the Petroleum Industry in Nigeria (EGASPIN) intervention value for mineral oil (petroleum hydrocarbon) of 5000 mg/kg [15]. By repeated application of CNB-Tech products, it is possible to meet a very strict regulatory standard for residual TPH level of less than 50 mg/kg. The changes in metal concentrations found in this study were attributed to (i) immobilization via chelate formation (ii) preferential supplementation of trace plant nutrient elements using the three products, (iii) natural electrochemical process whereby the positively or negatively charged organic molecules (generated during the natural transformation process occurring when the products were in use) bond with their counterparts in organic matter. These processes include oxidation, methylation, hydroxylation, carboxylation, coupling and polymerization [40] thereby enhancing bioavailability of the metals to microorganisms that utilize the organic matter supplied by the CNB –Tech products as energy source.

Microbial population found in a typical tropical soil under Nigerian climate is in the neighborhood of 8.19×10^6 cfu/mL [41]. Relative to this value, the population found in the contaminated OBM-DC (1.9 to 2.4×10^3 cfu/mL) showed suppressed microbial population, attributed to strong hydrocarbon (TPH level of 79, 200mg/kg) pollution. This is in agreement with the reports of [3]. The microbial population (1.45 to 3.15 x 10^7 cfu/mL) found in treated samples revealed restoration of soil microbial population using CNB-Tech products. It excelled over the value recorded in polluted material by over 7000 folds and higher than the value reported by [34], where TPH degraders and PAH degraders increased by one to two orders of magnitude via the addition of manure. Furthermore, the use of CNB-Tech products modified the pH value of the drilling wastes, transforming it from strongly alkaline (pH of 10) medium to pH of 7.90 medium; comparable to the 7.3±0.1 obtained by [34] for bioremediated soils. The very high pH of the untreated drilling waste materials could be attributed to some of the additives in the drilling fluid. Drilling fluids contain an internal phase of brine such as calcium salts [3]. This was confirmed by the high content of Ca (87 300 mg/kg) obtained in this study for the untreated material. One dose application of CNB-Tech products reduced this concentration by up to 62%, repeated dose application would definitely bring Ca level to any desired value.

Observations made during the recovery /fallow period were signs of drastic positive change in toxicity conditions, implying reduced toxicity. Reduction of soil toxicity by bioremediation, evidenced by increase in EC50 of the soil was reported by [34]. In this study, bioremediation using CNB-Tech products reduced toxicity in treated materials relative to untreated OBM-DC, evidenced by 100% positive effect on seedling germination potential and improved crop vegetative growth. Reduced material toxicity also explains the increased microbial activity of the treated matrices in comparison to the untreated drilling wastes, obtained in this study. The agricultural potential for the remediation end-products was also manifested by:

- increased microbial activities
- increased nitrogen-phosphorus-potassium (NPK) status
- increased soil crumby nature as against very viscous and pasty characteristics of untreated drilling wastes.

These nutrient elements (NPK) enhance microbial growth, microbial population, and microbial activity and consequently increase soil fertility [41]. By these, CNB-Tech products could overcome the extreme phytotoxicity [100% toxicity to seedling germination potential of maize and 100% inhibition to vegetative growth for three different types of plant (maize, fluted pumpkin and cassava)], caused by the untreated drilling waste. CNB-Tech products transformed oil-based drilling mud/cuttings to arable soil; capable of supporting seed germination and plant growth; excelling the performance of a control (farm soil apparently not impacted by drilling waste or crude oil) by 14%.

Electrical conductivity, a measure of dissolved ions in solution, is influenced by several soil physical and chemical properties such as salinity, saturation percentage, water content, bulk density, organic matter content, temperature and cation exchange capacity of the soil matrix. Impact of these influencing factors must be reflected in interpreting electrical conductivity effect on plant growth. Generally, elevated electrical conductivity and high salinity levels in agricultural soils may result in reduced plant growth and productivity or in extreme cases, the elimination of crops and native vegetation [42]. The reduction of electrical conductivity by 68% is a positive development because it demonstrates that the products could also modify the salinity of the material. In situations of very high initial electrical conductivity, there is a step-down CNB-Tech product as was carried out in this study and in situations of very low electrical conductivity, there is also a step-up CNB-Tech product as reported in a previous publication [30]. Results in this present study on excellent growth of crops planted in the remediated matrices were indicators of acceptable soil salinity level for plant growth. The beneficial use of the end-products obtained in this study for crop production were attributed to postulations based on findings from this study and previous works on this subject matter, which include:

- stimulation of beneficial microorganisms in soil, which enhances soil fertility [25]
- possible increased photosynthetic rate in plants evidenced by increased photosynthetic pigments (chlorophylls a and b) [40]
- increase in soil buffering capacity [28]
- increased soil moisture retention capacity by reducing hydrophobicity tendency [29]
- positive soil temperature modifications that enhance soil nutrient bioavailability to plants [31, 40]
- formation of stable chelates with toxic metals such as Pb, Cu and Cd in order to reduce their bioavailability to plants [40]
- preferential exclusion of the chelated toxic metals from soil

solution, allowing the plant nutrient elements to be assimilated into plant cells

- improvement of soil physicochemical properties via:
- increased aeration and water retention [29]
- activation of the macro and micro nutrients in soil in forms readily assimilated by plants [30, 40]
- improvement of plant root development and growth
- improvement of seed sprout of plants and subsequent shoot growth
- improved plant biomass production [26]
- enhanced soil nitrogen, phosphorus and potassium status for improved soil fertility
- acting as plant growth hormone, having positive stimulant action for plant growth [25, 26]
- improvement of soil permeability, promoting plant drought resistance [29]
- promotion of increased soil porosity and organic matter content, hence greatly promoting the microorganism activity and improving soil fertility.

Regarding leachate generation and management during the remediation exercise; fluid (leachate) produced as remediation progressed was recycled by incorporation into the biocell and used to regulate moisture content, thereby reducing water usage and conserving water resources. Expertise applied during the project ensured that at remediation project close-out, no isolated fluid system was actually produced. Nonetheless, the assessment of leachate effect on plant growth carried out in this work was to establish the fact that even in the event of accidental release of some fluid into the environment, there would be minimal risk to the receptor biotic community. More evaluations are still ongoing in this regard. Results from this study revealed that the leachate generated, though a concentrate, supported plant growth and when diluted with ordinary tap water gave a better support; reasons being that:

- toxic petroleum hydrocarbons in the contaminated drilling wastes have been destroyed to an acceptable level, evidenced by natural foamability of the concentrated leachate. Foamability would hardly occur if oil was still present

- leachate is also enriched with plant nutrients such as nitrogen, phosphorus and potassium

The process fluid, therefore, had some fertilizer value. The percentage decreases (1.50% and 23.45%) obtained for plant height and root length respectively, for the stock leachate was attributed to concentrated level of nutrients, confirmed by better performance of dilute leachate series. Naturally, in any formulated fertilizer, plant nutrients are applied at specified concentrations otherwise may hinder plant growth. Comparative evaluations of control system (soil treated with water only), stock leachate system (soil treated with leachate without any form of dilution) and systems with serial dilutions of the leachate (soil treated with leachate diluted with water by factors 1, 2, 3 and 4) revealed that the leachates were not toxic to receptor plants. The implication of this is that in the event of occasional spill of the leachate to the adjacent environment; dilution with water is, therefore, an adequate safety measure.

The ability of the end products to sustain the growth of green leafy vegetable: fluted pumpkin (*Telfairia ocidentis*) and root tuber crop, cassava (*Manihot esculenta Crantz*) and cereal crop (maize) is a demonstration of the utility of the remediation end product. It therefore stands that the use of CNB-Tech products as a biotechnological tool for hydrocarbon degradation in drilling waste converts these waste materials into non-toxic and potentially useful end products. In addition to the beneficial use of the remediation end-product for agricultural purposes, other possible utility options, shown in Figure 19, include:

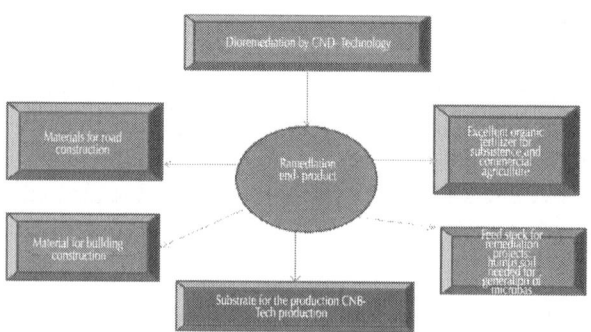

Figure 19: Potential utility of end - products from bioremediation using CNB-Tech products.

- material for road construction
- material for building construction
- substrate for the production CNB-Tech bioremediation agents
- excellent organic fertilizer for subsistence and commercial agriculture
- feedstock for bioremediation projects

Table 6 is a comparative evaluation of economic, operational and environmental implications of thermal technologies as reported by [3] and CNB-Technology based on the results and learning from this study.

Table 6: Comparative evaluation between thermal technology and CNB-technology

S/N	Thermal Technology	CNB-Tech
1.	Effective removal and recovery of hydrocarbons from solids	Effective removal of hydrocarbons from solid
2.	Possibility of recovering base fluid and end - product could be used for brick making	Effective recovery of free phase oil and end product has other uses apart from brick making
3.	Low potential for future liability	No future liability
4.	Requires short time	Time is relatively short
5.	High cost of handling environmental issues, since end- product dispersion would be below organic layer where vegetation growth is desired	Very minimized environmental issues
6.	Large volume of wastes is required to justify the cost of operation	Cost-effective for either small or large volume of wastes
7.	Requires tightly controlled process parameters	Does not require tightly controlled process parameters
8.	Heavy metals and salts are concentrated in processed solids	Reduces heavy metals and salts concentrations in process solid
9.	High operating temperatures can lead to safety risks	Low operating temperature. Operates at ambient temperature; modulation does not exceed 60°C.

10.	Requires several operators	Does not require several operators
11.	Process water contains some emulsified oils	Process water does not contain some emulsified oils
12.	Residue ash requires further treatment	No residue ash. End-product is clean soil
13.	End product is sterile and can no longer support plant Life	End product is fertile and can support microbial and plant Life

CONCLUSIONS AND RECOMMENDATIONS

This study revealed that it is possible to harness natural, biodegradable and local resource materials of Nigeria origin; translate them to scientifically formulated products that can be utilized for efficient biodegradation of hydrocarbon polluted matrices such as oil-based mud and drill cuttings within a reasonable short period of 6 days. This technology thus converts hydrocarbon polluted oil-based mud and drill cuttings to beneficial end-products of high order reuse such as soil amendment, without the generation of secondary waste materials. Field-scale trial adopting CNB-Technology is recommended.

ACKNOWLEDGEMENTS

This project was carried out under full financial support of the Remediation Department, Shell Petroleum Development Company (SPDC), Port Harcourt, Nigeria through the University Liaison Team of the company. The support of the Oil well Team of SPDC that facilitated the procurement of oil- based mud and drill cuttings is also acknowledged.

REFERENCES

1.   United Nations Environmental Programme (UNEP), 2011. Environmental Assessment of Ogoniland. P.1-262. ISBN:978-

92-807-9 Available on line at: http://postconflict.unep.ch/publications/OEA /UNEP_OEA.pdf

2.   Department of Health, Government of South Australia (DHGSA). Public Health Fact Sheet on Polycyclic Aromatic Hydrocarbons (PAHs): Health effects 2009 http://www.dh.sa.gov.au/pehs/PDF-files/ph-factsheet-PAHs-health.pdf

3.   Neff, M.M and Duxbury, MA. Composition, environmental fates, and biological effects of water based drilling muds and cuttings discharged to the marine environment: A Synthesis and Annotated Bibliography. Prepared for Petroleum Environmental Research Forum (PERF) and American Petroleum Institute. 2005. http://perf.org/pdf/APIPERFreport.pdf

4.   Gbadebo, A.M., Taiwo, A.M. and U. Eghele, U Environmental impacts of drilling mud and cutting wastes from the Igbokoda onshore oil wells, Southwestern Nigeria. Indian Journal of Science and Technology, 2010; 3(5), 504 -510.

5.   Environmental Protection Agency (EPA). An Assessment of the Environmental Implications of Oil and Gas Production: A Regional Case Study, 2008

6.   Osuji, L.C., Erondu, E.S and Ogali, R.E Upstream petroleum degradation of mangroves and intertidal shores: The Niger Delta Experience. Chemistry and Biodiversity, 2010: 7, 116 -128.

7.   Knez, D., Jerzy, A, G and Czekaj Trends in the drilling waste management. Acts Montanistica Rocnlk, 2006:11, 80-83.

8.   Morillon, A., Vidalie, J.F., hamzah, U.S., Suripno and Hadinota, E.K "Drilling and Waste management", SPE 73931, Intenationa; Conference on Health, Safety and Environment in oil and gas exploration and production, 2002: March 20-22

9.   Zimmerman, P.K. and Rober, J.D Oil-based drill cuttings treated by landfarming. Oil and Gas J, 1991: 12, 81-84

10.   Rojas-Avelizapa, N.G., Roldan-carrillo, T., Zegarra-Martinez, H., Munez-Colunga, A.M and Fernandez-Linares A field trial for an ext-situ bioremediation of a drilling mud-polluted site. Chemosphere 2007: 66, 1595-1600.

11.   Frydda, S and Randle, J.B Case study: Biological treatement of Geothermal drilling cuttings. Proceedings World Geothermal Congress, Bali, Indonesia, 25-29, 2010: 1-3.

12. Ouyang, W., Liu, H., Murygina, V., Yu, Y., Xiu, Z and Kalyuzhnyi, S Comparison of bio-augmentation and composting for remediation of oily sludge: A field-scale study in China. Process Biochemistry, 2005: 40, 3763 -3768.

13. Vidali, M. Bioremediation: An overview. Pure and Applied Chemistry, 2001: 73(7), 1163-1173

14. Jorgensen, K.S., Puutstinen, J and Suortt, A. –M Bioremediation of petroleum hydrocarbon-contaminated soil by composting in biopiles. Environmental Pollution, 2000: 107, 245-254.

15. Department of Petroleum Resources. Environmental Guidelines and Standard for the Petroleum Industry in Nigeria, 2002

16. Joel, O.F and Amajuoyi, C.A Determination of selected physicochemical parameters and heavy metals in a drilling cutting dump site at Ezeogwu–Owaza, Nigeria. J. Appl. Sci. Environ. Manage, 2009: 13(2), 27- 31.

17. Okparanma, R.N., Ayotamuno, J. M Polycyclic aromatic hydrocarbons in Nigerian oil-based drill-cuttings; evidence of petrogenic and pyrogenic effects. World Applied Sciences Journal 2010; 11 (4): 394-400, ISSN 1818-4952.

18. Nweke, C.O and Okpokwasili, G. C Drilling fluid base oil biodegradation potential of a soil Staphylococcus species. African Journal of Biotechnology 2003; 2 (9), pp. 293-295. http://www.academicjournals.org/AJB

19. Ayotamuno, J.M., Okparanma, R, N and Araka, P.P Bioaugmentation and composting of oil-field drill-cuttings containing polycyclic aromatic hydrocarbons (PAHs). Journal of Food, Agriculture & Environment 2009; l.7 (2): 6 5 8 - 664. www.world-food.net

20. Okparanma, R.N Ayotamuno, J.M and Araka, P.P Bioremediation of hydrocarbon contaminated-oil field drill-cuttings with bacterial isolates. African Journal of Environmental Science and Technology 2009 3 (5), pp. 131-140. Available online at http://www.academicjournals.org/AJEST

21. Ifeadi, C.N The treatment of drill cuttings using dispersion by chemical reaction (DCR). A paper prepared for presentation at the DPR Health, Safety & Environment (HSE) International Conference on Oil and Gas Industry in Port Harcourt, Nigeria. 2004.

22. Adekunle, I.M., Ajijo, M.R., Omoniyi, I.T and Adeofun, C.O Response of four phytoplankton species in some sections of Nigeria coastal waters to crude oil in controlled ecosystem. Int. J. Environ., Res., Iran, 2009; 4 (1): 65 -74 http://ijer.ut.ac.ir

23. Adekunle, I.M and Onianwa, P.C Functional group characteristics of humic acid and fulvic acid extracted from some agricultural wastes. Nigerian Journal of Science, Nigeria, 2001: 35 (1), 15 – 19.

24. Adekunle, I.M Evaluating environmental impact from utilization of bulk composted wastes of Nigerian origin using laboratory extraction test. Environmental Engineering and Management Journal 2010; 9 (5): 721 -729.: http://omicron.ch.tuiasi.ro/EEMJ/

25. Adekunle I.M., Adekunle, A.A., Akintokun, A.K., Akintokun, P and Arowolo, T.A Recycling of organic wastes through composting for land applications: a Nigerian experience. Waste Management & Research 2011; 29 (6): 582 – 593. DOI: 10.1177/ http://wmr.sagepub.com/content/29/6/582.abstract

26. Adekunle, I.M Bioremediation of soils contaminated with Nigerian petroleum products using composted municipal wastes. Bioremediation Journal, 2011; 15 (4): 230-241, DOI: 10.1080/10889868.2011.624137. http://dx.doi.org/10.1080/10889868.2011.624137

27. Adekunle I.M., Oguns, O., Shekwolo, P.D., Igbuku, O.O and Ogunkoya, O.O Assessment of population perception impact on value-added solid waste disposal in developing countries, a case study of Port Harcourt City, Nigeria. In: Xiao-Ying, Y (Ed) Municipal and Industrial Waste Disposal. Intech; 2012, p177-206.

28. Adekunle A. A., Adekunle, I.M., Igba, T. O Assessing the effect of bioremediation agent from local resource materials in Nigeria on soil pH. Journal of Emerging Trends in Engineering and Applied Sciences 2012; 3 (3) 526-532. http://jeteas.scholarlinkresearch.org/articles/Assessing%20the%20Effect%20of%20Bioremediation%20Agent.pdf

29. Adekunle A.A., I.M. Adekunle and Igba, T.O Impact of bioremediation formulation from Nigeria local resource materials on moisture contents for soils contaminated with petroleum products. International Journal of Engineering Research and

Development 2012; 2(4) 40-45 http://www.ijerd.com/paper/vol2-issue4/F02044045.pdf

30. Adekunle A.A, Adekunle, I.M. and Igba, T.O Assessing and forecasting the impact of bioremediation product derived from Nigeria local raw materials on electrical conductivity of soils contaminated with petroleum products. Journal of Applied technology in Environmental Sanitation 2012; 2 (1) 57 -66. http://www.trisanita.org/jates/atespaper2012/ates09v2n1y2012.pdf

31. Adekunle A.A., I. M. Adekunle and Igba T. O Soil temperature dynamics during bioremediation of petroleum products using remediation agent for Nigerian local resource materials. International Journal of Engineering Science and Technology 2012; 1 (4): 1-8. http://www.ijert.org/browse/june-2012-edition

32. Association of Official Analytical Chemists (AOAC), Official Method and Analysis of The Association oh The Official Analytical Chemists 11th Edition Washington D C, 1970.

33. Finar, I.L Organic Chemistry, volume I The Fundamental principles. 6th Ed, Longman, 1973.

34. Liu, W., Luo, Y and Teng, Y Bioremediation of oily sludge-contaminated soil by stimulating indigenous microbes. Environ Geochem health 2010: 32, 23 -29.

35. Ayotamuno, J.M., Okparanma, R.N., Davis, DD and allagoa, M. PAH removal from Nigerian oil-based drill-cuttings with spent oyster mushroom (Pleurotus ostretus) substrate. Journal of Food, Agriculture and Environment 2010: 8 (3 &4), 914 -919.

36. Rojas-Avelizapa, N.G., Roldan-Carrillo, T., Zegarra-Martinez, H., Munoz-Colunga, A.M and Fernadez-Linares A field trail for an ex-situ bioremediation of a drilling mud-polluted site. Chemospher, 2007: 66, 1595 – 1600.

37. Martin, J.A., Moreno, J.L., Hernandez, T and Garcia, C Bioremediation by composting of heavy oil refinery sludge in semiarid conditions. Biodegradation, 17:, 251 – 261.

38. Al-Nasrawi, H Biodegradation of Crude Oil by Fungi Isolated from Gulf of Mexico. J Bioremed Biodegrad 2012, 3:4

39. Mandal, A.K., Sarma, P. M., Singh, B., Jeyaseelan, C.P., Channasshettar, V.A., Lal, B and Datta, J bioremediation : an environment friendly sustainable biotechnological solution for

remediation of petroleum hydrocarbon contaminated waste. ARPN Journal of Science and Technology, 2012: 2, 1-12

40. Stevenson, F.J Humus Chemistry, 2004. Wiley & Sons

41. Obayori, O.S., Ilori, M.O., Adebusoye, S.A., Amund, O.O and Oyetibo, G.O Microbial population changes in tropical agricultural soil experimentally contaminated with crude petroleum. African Journal of Biotechnology, 2008: 7 (24), 4512-4520.

42. Corwin, D.L and Lesch, S.M. Apparent soil electrical conductivity measurements in agriculture. Computers and Electronics in Agriculture, 2005: 46, 11–4

Enhancement of Oxygen Mass Transfer and Gas Holdup Using Palm Oil in Stirred Tank Bioreactors with Xanthan Solutions as Simulated Viscous Fermentation Broths

Suhaila Mohd Sauid[1], Jagannathan Krishnan[1], Tan Huey Ling[1], and Murthy V. P. S. Veluri[2]

[1]Faculty of Chemical Engineering, Universiti Teknologi MARA (UiTM), 40450 Shah Alam, Selangor, Malaysia

[2]Faculty of Chemical Engineering, Manipal International University, 71800 Nilai, Negeri Sembilan Darul Khusus, Malaysia

ABSTRACT

Volumetric mass transfer coefficient ($k_L a$) is an important parameter in bioreactors handling viscous fermentations such as xanthan gum production, as it affects the reactor performance and productivity. Published literatures showed that adding an organic phase such as hydrocarbons or vegetable oil could increase the k_La. The present study opted for palm oil as the organic phase as it is plentiful in Malaysia. Experiments were carried out to study the effect of viscosity, gas holdup, and k_La on the xanthan solution with different palm oil fractions by varying the agitation rate and aeration rate in a 5 L bench-top bioreactor fitted with twin Rushton turbines. Results showed that 10% (v/v) of palm oil raised the k_La of xanthan solution by 1.5 to 3 folds with the highest k_La value of 84.44 h^{-1}. It was also found that palm oil increased the gas holdup and viscosity of the xanthan solution. The k_La values obtained as a function of power input, superficial gas velocity, and palm oil fraction were validated by two different empirical equations. Similarly, the gas holdup obtained as a function of power input and superficial gas velocity was validated by another empirical equation. All correlations were found to fit well with higher determination coefficients.

INTRODUCTION

In aerobic fermentations, the bioreactor performance greatly depends on its oxygen transfer capacities. Oxygen is a soluble substrate, but its solubility in aqueous media at ambient conditions is very low [1]. Thus, actively growing cells can consume all the dissolved oxygen quickly and, hence, oxygen has to be supplied continuously by mass transfer from air to the growth medium [2]. If the oxygen transfer rate to the aqueous phase exceeds the rate of oxygen consumed by the cells, cell growth continues at an exponential rate when other nutrients are not limited. However, when oxygen is not enough, the microorganisms' metabolic rate decreases drastically leading to reduced growth and productivity.

Xanthan gum is a natural polysaccharide produced by Xanthomonas campestris and is an important industrial biopolymer. It is widely used in industries such as foods and cosmetics, pharmaceutical and petroleum industry as emulsion stabilizer, thickener, dispersing agent and drilling fluid, and for many more applications [3]. Xanthan gum is soluble in cold or hot water and its solution can be highly viscous even at low concentration. Xanthan gum solutions show non-Newtonian behavior, that is, pseudoplastic or shear thinning, and the viscosity decreases with increasing shear rate. Besides, the viscosity also depends on its concentration, temperature, concentration of salts, and pH [4]. The fermentations of xanthan gum production are often associated with a significant decrease in oxygen transfer rate because of the increase in viscosity from accumulation of xanthan. As X. campestris is a strictly aerobic microorganism, oxygen is an essential nutrient both for growth and for xanthan production. Therefore, oxygen limitation turns into the controlling step in the whole process of xanthan production [5].

There have been various strategies to improve the oxygen transfer in bioreactors. Some of the previous researchers [6–10] adopted an approach of dispersing a nonaqueous, organic, second liquid phase that is immiscible to the system, referred to later as organic phase(s). The presence of organic phase modifies the medium in such a way that it could carry more oxygen and this approach was found successful in the past. The organic phase has strong affinity for oxygen so that it can increase the apparent solubility of oxygen in water [7]. The organic compounds used were hydrocarbons, perfluorocarbons, and vegetable oils.

This method also was applied in xanthan gum fermentation by Ju and Zhao [11] and Kuttuva et al. [12]. They postulated that this method has the ability to solve the viscosity problem and indirectly enhance the oxygen transfer. This was proven by Lo et al. [8] who found that in 3.5% of xanthan solution, the $k_L a$ values in the xanthan solution-hexadecane emulsion (0.50 v/v) system were higher than the $k_L a$ values obtained in a centrifugal packed bed reactor and stirred tank reactor. While those researchers opted for higher organic phase concentration, this study has focused on the effect of palm oil on the viscosity and the oxygen transfer characteristics in xanthan solution at lower oil concentrations.

In this work, palm oil was chosen as the organic phase to study the oxygen transfer characteristics such as effect of viscosity, gas holdup, and mass transfer coefficient on the xanthan gum solution by varying the agitation rate and aeration rate in a stirred tank bioreactor.

MATERIALS AND METHODS

Bioreactor

Experiments were carried out in an automated 5 L bench-top bioreactor (Biostat B, Sartorius BBI Systems, Germany) with a working volume of 4 L. The height/diameter ratio of glass vessel was 2 : 1. It was equipped with pH electrode (Mettler Toledo, Switzerland), dissolved oxygen (DO) probe (Mettler Toledo, Switzerland), temperature, and antifoam sensor. An overhead stirrer (Heidolph RZR 2102 Control, Germany) with agitation controller and torque reading was mounted on the stirrer shaft on top of the bioreactor for mixing. The bioreactor mixing system was equipped with four baffles and two impellers. Twin Rushton turbine blades spaced 80 mm apart having 64 mm diameter and 13 mm width were used. There was a ring sparger underneath the bottom turbine having of 14 holes 1 mm size each. Standard operating procedure was carried out on each experiment and the DO and pH probes were calibrated before starting bioreactor. For all experiments, the bioreactor was aerated at three different rates, namely, 0.25, 0.75, and 1.25 vvm and agitated at three different speeds, namely, 400, 600, and 800 rpm. The bioreactor used in this study is shown in Figure 1. All the experiments were carried out at atmospheric pressure and the temperature was maintained at 30°C.

Figure 1: Experimental setup.

Model Media

The investigation of the effect of palm oil on k_La was conducted in a model media, xanthan gum solutions to represent aqueous solutions of different viscosities as a fermentation broth. In this research, commercial food grade xanthan gum obtained from Bagus Bakery, Malaysia, was used and the xanthan gum solution was prepared at 0.25% (w/v). Xanthan gum was selected as it showed a good non-Newtonian (pseudoplastic) behavior as required besides being inexpensive.

Palm Oil

Palm oil used in this research was RBD (refined, bleached, and deodorized) palm olein obtained from Alami Technological Services Sdn Bhd, Malaysia. Its viscosity and density were measured as 67.85 cP and 818 kg/m³at the ambient temperature, respectively. The oxygen solubility of palm oil is 47.7 mg/L at 30°C [13] and it has low solubility in water, around 100 mg/L at 28°C [14].

Rheology Measurements

The viscosity measurement of xanthan gum solution was conducted using a viscometer (Brookfield LVDV-II+Pro, USA) at 30°C with a SC4-25 spindle. The shear stress versus shear rate data were analyzed as per the Ostwald de Waele or power law model given in (1). In order to study the effect of palm oil dosage, experiments were carried out at different volumetric fractions (0, 5, 10, 15, 20, and 50%) of palm oil in the xanthan solution:

$$\tau = K\gamma^{n}.$$

$$(1)$$

Power Input

The power input was measured by using an overhead stirrer with torquemeter (Heidolph RZR 2102 Control, Germany). The power input was obtained by applying the following:

$$P = 2\pi N \left(T - T_o\right).$$

(2)

Probe Response Time of DO Meter

In determining $k_L a$ values, it is important to find out the probe response time. Response time, τ_r, is defined as the time needed to record 63% of the final value measured when exposed to a stepwise change of concentration [15]. The probe response time was determined by transferring the dissolved oxygen probe (InPro 6820 Series, Mettler Toledo, Switzerland) from oxygen-free solution (0% saturation value) to an oxygen saturated solution (100% saturation value). The abrupt rise in the DO reading from 0% to 100% was monitored and recorded every five seconds. Then, the probe response was modeled as first-order dynamic as described by García-Ochoa et al. [5] in the following:

$$\frac{dC_{me}}{dt} = \frac{C_L - C_{me}}{\tau_r}.$$

(3)

Upon linearization, (3) became

$$\ln \frac{C_L - C_{me}}{C_L - C_{me_o}} = -\frac{t}{\tau_r}.$$

(4)

A plot of $\ln(C_L - C_{me} / L - C_{me_o})$ against t yielded a straight line with inverse slope value of $1/\tau_r$. From that the probe response time obtained was found to be 24.16 s.

Volumetric Mass Transfer Coefficient

For the $k_L a$ value determination the static gassing out method was employed throughout the experiments. This was performed by firstly purging the system (xanthan solution with palm oil) with nitrogen gas until the dissolved oxygen fell to zero. When DO was stabilized at 0% value, the nitrogen valve was switched off and, simultaneously, the aeration was started and time was marked as zero (t=0). The gradual increase of DO concentration was monitored and recorded until it reached a steady value at 100%. In order to calculate the became $k_L a$ value, (5) was used:

$$\frac{dC_L}{dt} = k_L a \left(C_L^* - C_L \right)$$

(5)

which on integration yielded

$$\ln \left(1 - \frac{C}{C^*} \right) = -k_L a \cdot t.$$

(6)

The $k_L a$ value was determined from the slope of plot $\ln(1 - C_L/C_L^*)$ versus t where C_L^* is the equilibrium dissolved oxygen concentration. The effect of the probe response was neglected as the time required for the oxygen transfer $1/k_L a$ was high compared to the dynamic response of the probe ($1/k_L a \gg 10\tau_r$) [12]. However, in case of the slower response, the dynamic response of the probemust be taken into account for the calculation of $k_L a$ [15–17] Considering the effect of response time of the DO probe used, the $k_L a$ value obtained from (6) served only as the initial guess value to compute the actual $k_L a$ value. Combining (3) and (5) yielded a nonlinear regression (7) for the experimental DO data which was later solved numerically to find the actual $k_L a$ value. Consider

$$C_{me} = C^* + \frac{C^* - C_{me_o}}{1 - \tau_r k_L a} \left[\tau_r k_L a \exp \left(-\frac{t}{\tau_r} \right) - \exp \left(-k_L a \cdot t \right) \right]$$

(7)

Gas Holdup

The gas holdup of the system was measured by the difference between the average liquid level with and without aeration using (8) given below. Consider

$$\varepsilon_G = \frac{H_{G+L} - H_L}{H_{G+L}}.$$

$$(8)$$

Empirical Correlations

There are several empirical correlations found in the published literature to predict $k_L a$ values; however, the one developed by Cooper et al. [18] is widely used. It relates $k_L a$ to the specific power input and the superficial gas velocity as given in the following:

$$k_L a = \delta \left(\frac{P_g}{V_L} \right)^{\beta} v_s^{\alpha}.$$

$$(9)$$

The power input term in the correlation includes all the effects of flow and turbulence on bubble dispersion and the mass transfer boundary layer according to Doran [1]. The values of the constants may vary depending on the system's geometry, the range of variables investigated, and the experimental technique applied. The range of values of β and α for Newtonian fluids varied between 0.4 to 0.95 and 0.2 to 0.75, respectively [19]

However, Nielsen et al. [20] modified (9) to include the organic phase term as in (10), where (P_g/V_L) represents the power input term, V_s represents the superficial gas velocity term, and ϕ_{ORG} represents the palm oil fraction term. The symbols of δ, β, α, and γ represent the empirical constants to be determined. Consider

$$k_L a = \delta \left(\frac{P_g}{V_L} \right)^{\beta} v_s^{\alpha} (1 - \phi_{ORG})^{\gamma}.$$

(10)

Gas holdup, ε_G, generally depends on superficial gas velocity, power consumption, surface tension and viscosity of liquids and solid concentration [15]. For an agitated reactor, the most common gas holdup correlation used is as shown by (11) [21–23]. The symbols χ, λ, and ω represent empirical constants specific to the systemunder investigation.Consider

$$\varepsilon_G = \chi \left(\frac{P_g}{V_L} \right)^{\lambda} v_s^{\omega}.$$

(11)

RESULTS AND DISCUSSION

The presence of palm oil in xanthan solution resulted in significant changes in the viscosity and rheological behavior, volumetric mass transfer coefficient, and gas holdup with respect to the different values of palm oil volume fraction, agitation rate, and aeration rate. The results obtained are discussed as follows.

Effect of Palm Oil on the Viscosity and Rheology of Xanthan Gum Solution

Xanthan solution (0.25%, w/v) exhibited a non-Newtonian pseudoplastic behavior. As palm oil was added into the xanthan solution at different oil fractions as presented in the Table 1, the rheological characteristics were changed. It was observed that the addition of palm oil slightly reduced the value of n and increased the value of K gradually in the power law (1). As the n value reduced with increase in the palm oil fractions, the degree of the pseudoplasticity of the xanthan solution increased with the increase in palm oil fraction than the pure xanthan solution.

Table 1: Rheological characteristics of xanthan solution at different palm oil fractions

Palm oil volume fraction (%)	Flow index, n	Consistency index, K (cPn)
0	0.4243	1108
5	0.4228	1191
10	0.4217	1280
15	0.4214	1334
20	0.4009	1399
50	0.344	2526

The viscosities of the xanthan solution with different palm oil fractions were measured as a function of the shear rate and the trend showed that, for all palm oil fractions, the viscosity decreased with the increase in shear rate and reached almost plateau at the higher shear rates. This trend showed that the solutions were in good agreement with the pseudoplastic nature of the xanthan solution. The apparent viscosity of the solutions at different palm oil fractions at a shear rate of $12.56\,s^{-1}$ was illustrated in Figure 2. It showed that the viscosity of xanthan solution increased gradually as the palm oil fraction was increased. Similar trend was obtained by Ma and Babosa-Cánovas [24] and they explained that, since the mean distance between the droplets was smaller, it facilitated compacting of oil, thereby leading to increase in the viscosity.

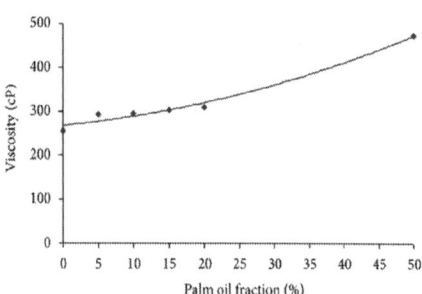

Figure 2: Variation of apparent viscosity of xanthan solution with palm oil fraction at a shear rate of $12.56\,s^{-1}$.

Effect of Palm Oil on k_La in Xanthan Gum Solution

Figure 3 showed the effect of palm oil on k_La as a function of agitation and aeration rate at different palm oil fraction. Comparing the values of k_La at the agitation rates of 400, 600, and 800 rpm, respectively, it was observed that the k_La values increased with the increase in agitation rate. Similar trends were also observed when the aeration rate was increased. The increase in aeration and agitation rate increased the degree of the liquid turbulence in the bioreactor. The liquid turbulence created was favorable to increase the k_La as it reduced the liquid film thickness at the gas-liquid interface [25]. Furthermore, the agitation was responsible for producing smaller bubbles, thus increasing the interfacial area, a, and it also increased the residence time of air bubbles [15, 25]. As a consequence, higher k_La with the increasing aeration and agitation rate was observed.

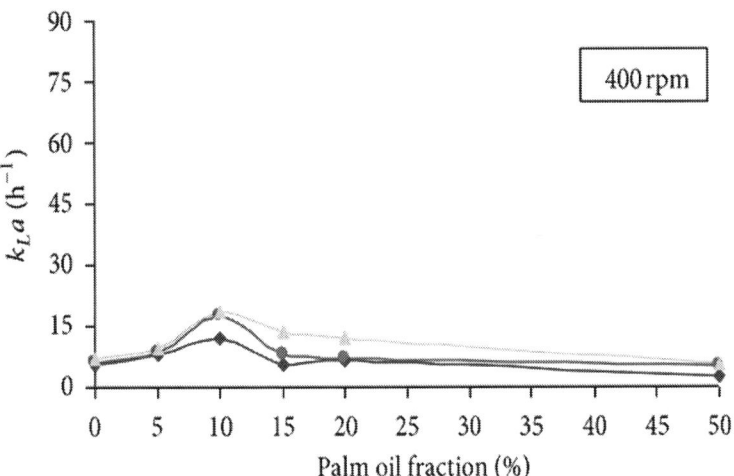

(a)

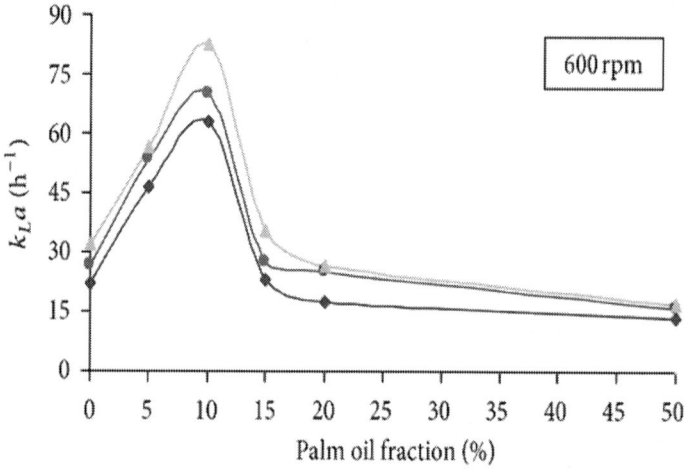

(b)

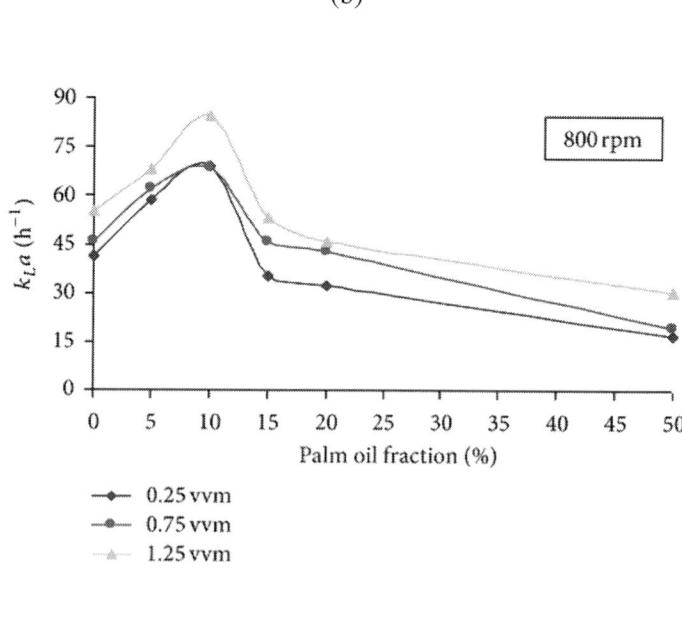

(c)

Figure 3: Influence of palm oil on k_La at varying agitation and aeration rate.

From the Figure 2, it is clearly showed that k_La was enhanced in the presence of palm oil. It was found that k_La was higher at 10% of oil fraction than that of pure xanthan solution for all the aeration and agitation rates. The highest k_La, 84.44 h^{-1}, was obtained at 1.25 vvm and 800 rpm. Most importantly, k_La values obtained for 10% oil fraction at 600 rpm were almost similar to that of 800 rpm. This observation proved that with the presence of 10% of palm oil in the xanthan solution, high oxygen transfer could be obtained without supplying any additional energy.

Rols et al. [26] suggested a few possible reasons for the increments in k_La with the addition of organic phases to the fermentation broth. The increments in k_La was resulted from the increase in the liquid turbulence contributed from the rigid organic phase droplets. In addition to that the organic phase formed a thin layer at the gas-liquid interface acting as intermediary for transport of oxygen to the aqueous phase.

Further, as suggested by Yoshida et al., [6] the spreading coefficient value, S, also plays a major role in altering the oxygen transfer capability of the system. For an organic phase having a negative value of spreading coefficient wouldform into floating lens like droplets in the system, while the coefficient being positive, the organic phase would spread on the water surface like a surface active agent to lower the surface tension and thereby increase the interfacial area, a, ultimately increasing k_L a. Since the palm oil used in the present work also has a positive value of spreading coefficient ($S = 38.3$ mN/m), it is evident that the surface tension of the xanthan solution was reduced by the palm oil and hence decreased the air bubble size leading to increased interfacial area and k_L a. Similar results were reported by Yoshida et al. [6] for toluene in water and oleic acid in water system with toluene and oleic acid having positive values of spreading coefficient.

The enhancements of k_La at 10% of oil fraction were more evident at 400 rpm and 600 rpm than the enhancements at 800 rpm. This could be seen in the form of enhancement factor as shown in Table 2. Enhancement factor can be defined as the ratio of k_La value in xanthan solution with palm oil, $(k_La)_{po}$ to that of pure xanthan solution, $(k_La)_o$ both measured at the same aeration and agitation rate. In this table, it could be observed that k_La was enhanced by almost three times at 400 rpm and 600 rpm compared to that of 800 rpm as the enhancement factor was only 1.57 in average.

Table 2: The enhancement factor of $k_L a$ at 10% palm oil fraction

Agitation rate	Aeration rate		
	0.25 vvm	0.75 vvm	1.25 vvm
	(kLa)po/ (kLo)o	(kLa)po/ (kLo)o	(kLa)po/ (kLo)o
400 rpm	2.23	2.95	2.71
600 rpm	2.85	2.65	2.59
800 rpm	1.68	1.49	1.52

The enhancement factor decreased with increase in the aeration and agitation rate. Similar trend was also observed by Galaction et al. [9] for the dodecane as the organic phase in the fermentation broth. They justified that reduction in the enhancement factor was due to the disruption of the dodecane superficial film or the removal of dodecane droplets from the bubble surface caused by the intensification of the mixing and turbulence at higher aeration and agitation rate. However, in this study, when the agitation rate was increased to 800 rpm, the mixing was intense and with further addition of the palm oil yielded tiny bubbles, thereby lowering the value of k_L [1] which contributed to the lower enhancement factor at 800 rpm compared to that of 400 and 600 rpm.

However, further increase in the oil fractions was found to decrease the $k_L a$ values. With the increase in palm oil fraction, the viscosity as well as the degree of the pseudoplasticity was found to increase as shown in Table1. Small bubbles that produced in such solution remained in the solutions due to reduced rising velocity. They became rigid spheres having lower $k_L a$ value due to surface immobility and no gas circulation [17, 28]. Therefore, it is concluded that the effect of the increase in viscosity and the change in rheology of the solution with the increase in palm oil fraction outweighed the effect of decreased surface tension, which resulted in the decrease of $k_L a$.

Effect of Palm Oil on Gas Holdup in Xanthan Gum Solution

As seen in Figure 4, the effect of palm oil on gas holdup in the xanthan solution showed similar trend for all the ranges of agitation and aeration rate studied. The gas holdup increased with the increase in palm oil fraction up to 15%. The increase in the gas holdup with the palm oil addition could be due to the dispersion of small bubbles in the system. As discussed earlier in Section 3.2, palm oil might have spread on the bubble surface, thus reducing the surface tension and decreasing the bubble size. The smaller bubbles had induced the gas holdup due to their low rise velocity and longer residence time in the system than the bigger bubbles [1,29]. This result is in agreement with Kawase and Moo-Young [30] who also found increments in gas holdup in their CMC solution with antifoam addition. They postulated that antifoam promoted the bubble breakup rather than bubble coalescence.

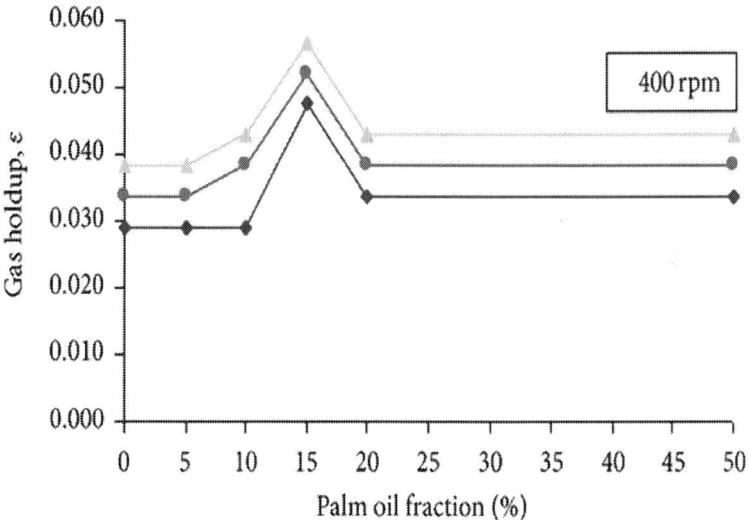

(a)

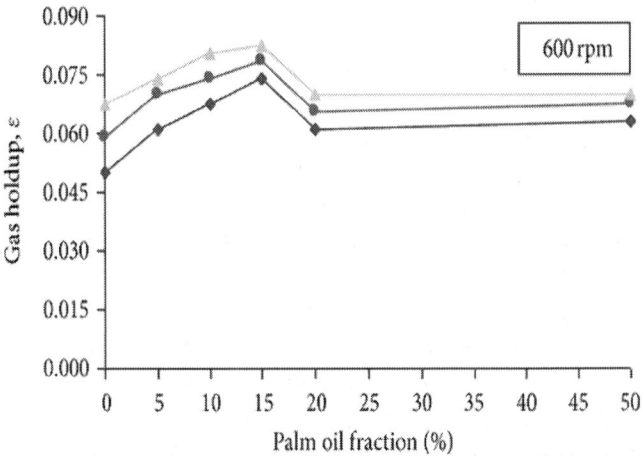

(b)

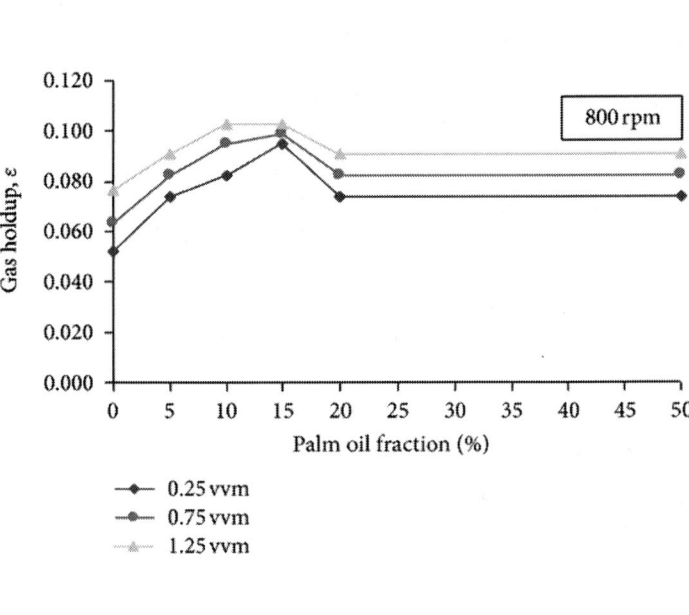

(c)

Figure 4: Influence of palm oil on gas holdup at varying agitation and aeration rate.

However, when the palm oil fraction reached beyond 20%, the gas holdup decreased and maintained the same until 50%. As mentioned in Section 3.1, palm oil addition into the xanthan solution changed the viscosity and the degree of pseudoplasticity of the solution. According to Machon et al. [31], the increase in the degree of pseudoplasticity had affected the stirrer's ability to dissipate power to the system to create smaller bubbles when the sparged air passed through and into the stirrer region. In addition, when the viscosity of the system increased, the degree of the liquid turbulence decreased which induced the bubble coalescence and these occurrences increased the proportion of larger bubbles in the liquid [32, 33] which had high rise velocities raced to the surface and ultimately reduced the gas holdup. Furthermore, even though the gas holdups at these fractions (20% and 50%) were lower than at the rest of the oil fraction, they were still higher than the holdup in pure xanthan solution. Hence, some of the tiny bubbles tend to remain lodged in the solution which contributed to the higher gas holdup. Similar occurrences were also observed by Doran [1].

Correlations for Volumetric Mass Transfer Coefficient in Xanthan Gum Solution

This experimental study is the first of its kind to report the correlations for $k_L a$ obtained for palm oil in xanthan solution. Other organic phases were used in various fermentation broths and for a few bioreactor systems. Equation (10) was used to fit the correlations for $k_L a$ in the literature which are listed in Table 3. whereas in this study, the experimental data was fitted into both correlations (9) and (10) to find $k_L a$ for palm oil in xanthan solution. Equation (9) considered the effect of superficial gas velocity and power input on $k_L a$ while (10) took into account the effect of palm oil fraction in addition. The experimental data was fitted well into both equations with high correlation coefficients. The values of the empirical constants obtained are listed in Table 3.

Table 3: The empirical constants for k_La reported in the literature and in the current study

Reference	Liquid system	Type of impeller	Constant	Constant	Constant	Constant	Valid for (kW/m3)	R2	Average error, %	Correlation
[10]	Perfluorocarbon (PFC) in YPD medium with inactive cells	Rushton turbine	0.302	0.699	-1.378	NA	NA	15.7	Nielsen et al. [20] (10)
[20]	n-Hexadecane in broth medium	Rushton turbine	0.31	0.70	1.70	NA	NA	NA	
[27]	Methyl ricinoleate in Yarrowia lipolytica broth	Rushton turbine	0.6	0.8	-22	NA	NA	NA	
[27]	Tween 80 in Yarrowia lipolytica broth	Rushton turbine	0.4	0.7	-449	NA	NA	NA	
Current study	Xanthan gum solution with 0 to 10% palm oil fraction	Rushton turbine	0.4773	0.1620	-6.6029	0.079 to 1.11	0.77	58.75	Nielsen et al. [20] (10)
	Xanthan gum solution with 0% palm oil fraction	Rushton turbine	0.6674	0.2076	-6.6029	0.141 to 1.11	0.89	46.55	
Current study	Xanthan gum solution with 5% palm oil fraction	Rushton turbine	0.5361	0.1225	-6.1622	0.094 to 0.89	0.78	58.74	
	Xanthan gum solution with 10% palm oil fraction	Rushton turbine	0.4078	0.1702	-8.2896	0.079 to 0.89	0.68	46.47	
	Xanthan gum solution with 15% palm oil fraction	Rushton turbine	0.4401	0.4253	-2.7865	0.058 to 0.83	0.74	40.93	Nielsen et al. [20] (10)
	Xanthan gum solution with 20% palm oil fraction	Rushton turbine	0.4885	0.2777	-1.6270	0.039 to 0.72	0.98	15.59	
	Xanthan gum solution with 50% palm oil fraction	Rushton turbine	0.4334	0.1605	-0.0697	0.031 to 0.64	0.86	34.40	

Current study	Xanthan gum solution with 0% palm oil fraction	Rushton turbine	0.6674	0.2076	—	0.141 to 1.11	0.89	46.55	Cooper et al. [18] (9)
	Xanthan gum solution with 5% palm oil fraction	Rushton turbine	0.5361	0.1225	—	0.094 to 0.89	0.78	58.73	
	Xanthan gum solution with 10% palm oil fraction	Rushton turbine	0.4078	0.1702	—	0.079 to 0.89	0.68	46.47	
	Xanthan gum solution with 15% palm oil fraction	Rushton turbine	0.5343	0.3054	—	0.058 to 0.83	0.96	18.80	
	Xanthan gum solution with 20% palm oil fraction	Rushton turbine	0.4994	0.2740	—	0.039 to 0.72	0.98	14.00	
	Xanthan gum solution with 50% palm oil fraction	Rushton turbine	0.4458	0.3391	—	0.031 to 0.64	0.91	27.67	

While using (9), the k_La values obtained were highly dependent on the specific power input supplied and the empirical constants obtained were within the range suggested by Winkler [19]. This indicated that a change in specific power input would change the k_La significantly than the change in the superficial gas velocity. However, while using (10), the exponent of palm oil fraction dominated the correlation. This meant that the oil fraction had the highest influence on k_La. The negative sign at the exponent value showed that adding palm oil up to 10% volume fraction would increase the k_La.

According to Kawase and Moo-Young [34], for a non- Newtonian system, the suggested values of β are between 0.37 and 0.80, while, for α, the values are between 0.20 and 0.84. Comparing the values in Table 3, it was found that β values obtained in this study were within the range specified by Kawase and Moo-Young [34]. However, for α, the values were found lower at 5% and 10% oil fractions, respectively. As this constant is a measure of aeration rate, it corresponds to lower kLa values compared to that of other oil fractions. It might be due to insufficient aeration compared to the other oil fraction. However, the constant γ was not comparable. This couldbedue tothedifference inthe bioreactor geometry, type of organic phase used in the study, and different rheological behavior of the bioreactor systems

Correlations for Gas Holdup in Xanthan Gum Solution

Again this study is the first of its kind to report the correlation for obtained for palm oil in xanthan solution. Few researchers used (11) for their agitated vessels handling different solution, to fit the correlation for which are listed in Table 4.

Table 4: The empirical constants for ε_G by (11) reported in the literature and in the current study

Reference	Liquid system	Constant	Constant	Constant	Type of impeller	R2	Average relative error (%)	Valid for (kW/m3)
[21]	Water	—	0.25	0.75	Flat blade turbine	NA	NA	NA
[22]	Na2SO4	—	0.4903	0.5788	Rushton turbine	NA	9.3	NA
[23]	Tap water	—	0.478	0.4910	Disc turbine-pitched blade	0.98	NA	NA
[23]	Tap water	—	0.4244	0.6904	Rushton turbine	0.98	NA	NA
	Xanthan gum solution with 5% palm oil fraction	0.0061	0.3501	0.1435	Rushton turbine	0.90	12.15	0.094–0.89
	Xanthan gum solution with 10% palm oil fraction	0.0065	0.3528	0.1626	Rushton turbine	0.93	10.87	0.079–0.89
	Xanthan gum solution with 15% palm oil fraction	0.0161	0.2507	0.0773	Rushton turbine	0.98	3.27	0.058–0.83
	Xanthan gum solution with 20% palm oil fraction	0.0119	0.2558	0.1395	Rushton turbine	0.96	5.99	0.039–0.72
	Xanthan gum solution with 50% palm oil fraction	0.0178	0.1890	0.1690	Rushton turbine	0.84	12.06	0.031–0.64

The correlations found in the literature considered only the empirical constants λ and ω in (11), whereas in this study all the three terms were considered and the results are listed in Table 4. For xanthan solution, the empirical constants ranged within $0.0061 < \chi < 0.0178$, $0.1890 < \lambda < 0.3528$, and $0.0773 < \omega < 0.2180$. As the value of λ was more than the other two constants for all palm oil fractions, it could be concluded that the change in specific power input had more impact on gas holdup than the change in superficial gas velocity for xanthan solution.

As shown in Table 4, it can be seen that the λ values obtained in this study are comparable with the value obtained by de Figueiredo and Calderbank [21]. However, due to the significant difference in the type and properties such as viscosity and rheological behavior of the organic phase used, a slight variation in the values of λ and ω is observed when compared to the literature.

CONCLUSIONS

The volumetric mass transfer coefficient, $k_L a$, of xanthan solution was measured at varying operating variables and palm oil fraction in a stirred tank bioreactor. It was evident that the addition of palm oil up to 10% volume fraction enhanced $k_L a$ by 1.5 to 3 folds with the highest $k_L a$ value of $84.44\ h^{-1}$. It was also found that increase in palm oil fraction increased the viscosity and rheology of the xanthan solution. This favorable effect was contributed from the properties of palm oil that promoted the production of small bubbles. This ultimately outweighed the positive effect on bubble size and therefore affected both and gas holdup. The $k_L a$ values obtained were also correlated with the power input, superficial gas velocity, and palm oil fraction in two different forms of equations which were found to fit well with very high correlation coefficients.

AUTHORS' CONTRIBUTION

All of the authors contributed to a similar extent, overall, and all authors have seen and agreed to the submission of this paper.

ACKNOWLEDGMENTS

This work was financially supported by the Ministry of Science, Technology and Innovation (MOSTI), Malaysia, through the research Grant no. 100-IRDC/SF 16/6/2(50/2007).

REFERENCES

1. P. M. Doran, Bioprocess Engineering Principles, Academic Press, London, UK, 1996.

2. S. M. Sauid and V. V. P. S. Murthy, "Effect of palm oil on oxygen transfer in a stirred tank bioreactor," Journal of Applied Sciences, vol. 10, no. 21, pp. 2745–2747, 2010.

3. B. Katzbauer, "Properties and applications of xanthan gum," Polymer Degradation and Stability, vol. 59, no. 1–3, pp. 81–84, 1998.

4. F. García-Ochoa, V. E. Santos, J. A. Casas, and E. Gómez, "Xanthan gum: production, recovery, and properties," Biotechnology Advances, vol. 18, no. 7, pp. 549–579, 2000.

5. F. García-Ochoa, E. G. Castro, and V. E. Santos, "Oxygen transfer and uptake rates during xanthan gum production," Enzyme and Microbial Technology, vol. 27, no. 9, pp. 680–690, 2000.

6. F. Yoshida, T. Yamane, and Y. Miyamoto, "Oxygen absorption into oil-in-water emulsions: a study on hydrocarbon fermentors," Industrial and Engineering Chemistry, vol. 9, no. 4, pp. 570–577, 1970.

7. J. L. Rols and G. Goma, "Enhanced oxygen transfer rates in fermentation using soybean oil-in-water dispersions," Biotechnology Letters, vol. 13, no. 1, pp. 7–12, 1991.

8. Y. M. Lo, C. H. Hsu, S. T. Yang, and D. B. Min, "Oxygen transfer characteristics of a centrifugal, packed-bed reactor during viscous xanthan fermentation," Bioprocess and Biosystems Engineering, vol. 24, no. 3, pp. 187–193, 2001.

9. A.-I. Galaction, D. Cascaval, C. Oniscu, and M. Turnea, "Enhancement of oxygen mass transfer in stirred bioreactors using oxygen-vectors. 1. Simulated fermentation broths," Bioprocess and Biosystems Engineering, vol. 26, no. 4, pp. 231–238, 2004.

10. P. F. F. Amaral, M. G. Freire, M. H. M. Rocha-Leão, I. M. Marrucho, J. A. P. Coutinho, and M. A. Z. Coelho, "Optimization of oxygen mass transfer in a multiphase bioreactor with perfluorodecalin as a second liquid phase," Biotechnology and Bioengineering, vol. 99, no. 3, pp. 588–598, 2008.

11. L.-K. Ju and S. Zhao, "Xanthan fermentations in water/oil dispersions," Biotechnology Techniques, vol. 7, no. 7, pp. 463–468, 1993.

12. S. G. Kuttuva, A. S. Restrepo, and L.-K. Ju, "Evaluation of different organic phases for water-in-oil xanthan fermentation," Applied Microbiology and Biotechnology, vol. 64, no. 3, pp. 340–345, 2004.

13. J. C. Allen and R. J. Hamilton, Rancidity in Foods, Chapman & Hall, New York, NY, USA, 1994.

14. K. Ahmad, C. C. Ho, W. K. Fong, and D. Toji, "Properties of palm oil-in-water emulsions stabilized by nonionic emulsifiers," Journal of Colloid and Interface Science, vol. 181, no. 2, pp. 595–604, 1996.

15. P. F. Stanbury and A. Whitaker, Principles of Fermentation Technology, Butterworth-Heinemann, Great Britain, UK, 2nd edition, 1995.

16. W. A. Brown, "Developing the best correlation for estimating the transfer of oxygen from air to water,"Chemical Engineering Education, vol. 35, no. 2, pp. 134–147, 2001.

17. F. Garcia-Ochoa and E. Gomez, "Bioreactor scale-up and oxygen transfer rate in microbial processes: an overview," Biotechnology Advances, vol. 27, no. 2, pp. 153–176, 2009.

18. C. M. Cooper, G. A. Fernstrom, and S. A. Miller, "Performance of agitated gas-liquid contactors,"Industrial & Engineering Chemistry, vol. 36, pp. 504–509, 1944.

19. M. A. Winkler, Ed., Chemical Engineering Problems in Biotechnology, Elsevier Science, England, UK, 1990.

20. D. R. Nielsen, A. J. Daugulis, and P. J. McLellan, "A novel method of simulating oxygen mass transfer in two-phase partitioning bioreactors," Biotechnology and Bioengineering, vol. 83, no. 6, pp. 735–742, 2003

21. M. M. L. de Figueiredo and P. H. Calderbank, "The scale-up of aerated mixing vessels for specified oxygen dissolution rates," Chemical Engineering Science, vol. 34, no. 11, pp. 1333–1338, 1979.

22. T. Moucha, V. Linek, and E. Prokopová, "Gas hold-up, mixing time and gas-liquid volumetric mass transfer coefficient of various multiple-impeller configurations: rushton turbine, pitched blade and techmix impeller and their combinations," Chemical Engineering Science, vol. 58, no. 9, pp. 1839–1846, 2003.

23. K. Saravanan, V. Ramamurthy, and K. Chandramohan, "Gas holdup in multiple impeller agitated vessels," Modern Applied Science, vol. 3, no. 2, pp. 49–59, 2009.

24. L. Ma and G. V. Barbosa-Cánovas, "Rheological characterization of mayonnaise. Part II: flow and viscoelastic properties at different oil and xanthan gum concentrations," Journal of Food Engineering, vol. 25, no. 3, pp. 409–425, 1995.

25. R. K. Finn, "Agitation-aeration in the laboratory and in industry," Microbiology and Molecular Biology Reviews, vol. 18, pp. 254–274, 1954.

26. J. L. Rols, J. S. Condoret, C. Fonade, and G. Goma, "Mechanism of enhanced oxygen transfer in fermentation using emulsified oxygen-vectors," Biotechnology and Bioengineering, vol. 35, no. 4, pp. 427–435, 1990.

27. N. Gomes, M. Aguedo, J. Teixeira, and I. Belo, "Oxygen mass transfer in a biphasic medium: influence on the biotransformation of methyl ricinoleate into γ-decalactone by the yeast Yarrowia lipolytica," Biochemical Engineering Journal, vol. 35, no. 3, pp. 380–386, 2007.

28. K. G. Clarke and L. D. C. Correia, "Oxygen transfer in hydrocarbon-aqueous dispersions and its applicability to alkane bioprocesses: a review," Biochemical Engineering Journal, vol. 39, no. 3, pp. 405–429, 2008.

29. G. D. Najafpour, Biochemical Engineering and Biotechnology, Elsevier, Amsterdam, The Netherlands, 2007.

30. Y. Kawase and M. Moo-Young, "Influence of antifoam agents on gas hold-up and mass transfer in bubble columns with non-newtonian fluids," Applied Microbiology and Biotechnology, vol. 27, no. 2, pp. 159–167, 1987.

31. V. Machon, J. Vlcek, A. W. Nienow, and J. Solomon, "Some effects of pseudoplasticity on hold-up in aerated, agitated vessels," The Chemical Engineering Journal, vol. 19, no. 1, pp. 67–74, 1980.

32. S. J. Arjunwadkar, K. Sarvanan, P. R. Kulkarni, and A. B. Pandit, "Gas-liquid mass transfer in dual impeller bioreactor," Biochemical Engineering Journal, vol. 1, no. 2, pp. 99–106, 1998.

33. F. García-Ochoa and E. Gómez, "Mass transfer coefficient in stirred tank reactors for xanthan gum solutions," Biochemical Engineering Journal, vol. 1, no. 1, pp. 1–10, 1998.

34. Y. Kawase and M. Moo-Young, "Volumetric mass transfer coefficients in aerated stirred tank reactors with Newtonian and non-Newtonian media," Chemical Engineering Research and Design, vol. 66, no. 3, pp. 284–288, 1988.

Citations

CHAPTER 1

Faraz Shah and Ilia G. Polushin, "Design of Telerobotic Drilling Control System with Haptic Feedback," Journal of Control Science and Engineering, vol. 2013, Article ID 901610, 15 pages, 2013. doi:10.1155/2013/901610.

CHAPTER 2

Wang Yu, Yao Jianyi, and Li Zhijun, "Design and Development of Turbodrill Blade Used in Crystallized Section," The Scientific World Journal, vol. 2014, Article ID 682963, 12 pages, 2014. doi:10.1155/2014/682963.

CHAPTER 3

Chuanliang Yan, Jingen Deng, and Baohua Yu, "Wellbore Stability in Oil and Gas Drilling with Chemical-Mechanical Coupling," The Scientific World Journal, vol. 2013, Article ID 720271, 9 pages, 2013. doi:10.1155/2013/720271.

CHAPTER 4

Nripen Mondal, Biswajit Sing Sardar, Ranendra Nath Halder, and Santanu Das, "Observation of Drilling Burr and Finding out the Condition for Minimum Burr Formation," International Journal of Manufacturing Engineering, vol. 2014, Article ID 208293, 12 pages, 2014. doi:10.1155/2014/208293.

CHAPTER 5

Pingting Liu, Zhiyu Huang, Hao Deng, Rongsha Wang, and Shuixiang Xie, "Synthesis and Performance Evaluation of a New Deoiling Agent for Treatment of Waste Oil-Based Drilling Fluids," The Scientific World Journal, vol. 2014, Article ID 852503, 9 pages, 2014. doi:10.1155/2014/852503.

CHAPTER 6

Junyi Liu, Zhengsong Qiu, Wei'an Huang, Dingding Song, and Dan Bao, "Preparation and Characterization of Latex Particles as Potential Physical Shale Stabilizer in Water-Based Drilling Fluids," The Scientific World Journal, vol. 2014, Article ID 895678, 8 pages, 2014. doi:10.1155/2014/895678.

CHAPTER 7

Hwee Ling Lim, "Assessing Level and Effectiveness of Corrosion Education in the UAE," International Journal of Corrosion, vol. 2012, Article ID 785701, 10 pages, 2012. doi:10.1155/2012/785701.

CHAPTER 8

Wei Yu and Huaqing Xie, "A Review on Nanofluids: Preparation, Stability Mechanisms, and Applications," Journal of Nanomaterials, vol. 2012, Article ID 435873, 17 pages, 2012. doi:10.1155/2012/435873.

CHAPTER 9

Iheoma M. Adekunle, Augustine O. O. Igbuku, Oke Oguns and Philip D. Shekwolo (2013). Emerging Trend in Natural Resource Utilization for Bioremediation of Oil — Based Drilling Wastes in Nigeria, Biodegradation - Engineering and Technology, Dr. Rolando Chamy (Ed.), ISBN: 978-953-51-1153-5, InTech, DOI: 10.5772/56526.

CHAPTER 10

Suhaila Mohd Sauid, Jagannathan Krishnan, Tan Huey Ling, and Murthy V. P. S. Veluri, "Enhancement of Oxygen Mass Transfer and Gas Holdup Using Palm Oil in Stirred Tank Bioreactors with Xanthan Solutions as Simulated Viscous Fermentation Broths," BioMed Research International, vol. 2013, Article ID 409675, 9 pages, 2013. doi:10.1155/2013/409675.

Index